The Complete Manual of Wood Bending

The Complete Manual of Wood Bending

MILLED, LAMINATED, AND STEAM-BENT WORK

by

LON SCHLEINING

Linden Publishing Inc.
Fresno CA

The Complete Manual of Wood Bending
by
Lon Schleining
Drawings by James Goold
ISBN: 0-941936-54-6

First printing: January 2002
Printed in: The United States of America
Library of Congress Cataloging-in-Publication Data:

Schleining, Lon
The complete manual of wood bending / by Lon Schleining.
p. cm.
Includes index.
ISBN0-941936-54-6 (pbk.)
1. Wood bending. I. Title.

TS852 .S38 2001
684'.08--dc21 2001050756

Your safety is your responsibility. Neither the author nor the publisher assumes any responsibility for any injuries suffered or for damages or other losses incurred that may result from material presented in this publication.

LINDEN PUBLISHING

The Woodworker's® Library

**Linden Publishing Inc.
2006 S. Mary
Fresno CA 93721
800-345-4447
www.lindenpub.com**

The Complete Manual of Wood Bending

CONTENTS

Dedication

FOR MY FRIEND AND MENTOR, the late Carl Shafer. Absent his quiet inspiration, you would not be holding this book in your hands.

Acknowledgments

WHEN PRODUCING A BOOK, editors, designers, and publishers usually top the thank-you list. In this case, many people not singled out by name have made contributions. They are the hundreds of participants in my woodworking seminars who ask penetrating questions and offer unique solutions to problems. They are clients who endlessly challenge my skills with imaginative and seemingly impossible requests. I'd like to thank them all for keeping me on my toes and encouraging me to push the woodworking envelope.

During the decade we worked together, my long-time friend Jeff Smith played a significant role in helping to devise many of the jigs, fixtures, and techniques you see on these pages.

Joe Romero tirelessly took many of the photographs in the book, including the one on the cover. Lee Stocker introduced me to tungsten film, and deserves much of the credit for the degree to which I satisfactorily produced the balance of the shots.

Julie Schleining, always supportive and mindful of the big picture (the bills have to be paid every month), gave me the room to pursue projects like this book despite the fact that it took me away from more immediate and more pressing concerns.

Editor Laura Tringali took a very rough manuscript and artfully sculpted it into a book. Amazingly, she managed to translate my obscure woodworking terminology into the English language as well. Artist Jim Goold produced crisp, precise drawings.

Richard Sorsky took a chance on an untested author. I appreciate his patience in the face of my endless requests for more time to work on the book. Without his persistence, no doubt I'd still be tinkering with it.

Introduction

TO THE UNINITIATED, the processes of curving wood can seem mysterious indeed. To the experienced, the techniques are surprisingly straightforward. This book will help you to join the ranks of those in the know. Once you understand the basic principles, become familiar with a few rules of thumb, and do a bit of carefully guided experimentation, you'll likely become the curved woodworking wizard of your neighborhood.

Curves are usually made in one of three ways: by milling solid lumber into a curved shape; by lamination-bending, where thin strips of wood are glued together around a form; and by steam-bending, where solid wood is heated with steam and bent. There is really no mystery to any of these methods. Milling a perfect circle is not that involved with the right setup. Lamination-bending is nothing more than cutting some strips and sticking them together with adhesive. Steam-bending isn't black magic, it is only a matter of controlling some variables. Using one of these methods (or perhaps a combination of methods), just about any curved shape can be built out of wood.

In order to decide how to go about making a curved component, you need to know some of the realities of the three basic curve-making techniques. You'll learn about that in this book, and while we're at it, we'll explode some of the myths about the methods in an effort to help you decide the best way to build your particular project. Concentrating on power-tool operations but including discussions of hand-tool work, you'll learn about every important part of the process, from design and layout to building shop gizmos and handling tools and materials safely.

Can novices become proficient at bending wood? While the techniques described may be a stretch for beginning woodworkers, they should be well within the capabilities of those with an understanding of basic woodworking. The assumption is that if you are about to tackle a curved project, you have a solid grounding in woodworking fundamentals. From there you will proceed to learn as much as you can about the subject, make some decisions, and carefully plan your strategy.

The learning is key. Most woodworkers find working with squares and rectangles quite familiar, since most woodworking machines are built to make boards perfectly straight or flat. But when it comes to making curves, it's a different story. Some of the materials are different. Tools may be used in unfamiliar ways. Many techniques may be new, since, depending on the type of bending, you very well could be using procedures more common to boatbuilding, stairbuilding, or metalworking.

Of course, once you try the techniques and tooling that are described here, it's likely you'll discover modifications. But whether you start from scratch, have experience with bending, or are building a project beyond the scope of this book, you will soon find that the principles covered here have broad application for both prototypes and small production runs.

WHAT YOU'LL FIND IN THIS BOOK

The format of the book facilitates problem-solving in a way that lends itself to shop use. Nothing would please me more than to see a glue-stained copy of this book stuffed onto an already crowded shelf between the drill bits and sandpaper. A glossary provides definitions should you get lost in the vocabulary of curved woodworking. A Quick Guide outlines the information in abbreviated form, to help you access the information you need—fast. The main section of the book starts with design, planning, and drawing, and continues through the techniques for making curves and the practical applications for the techniques.

Drawing is nothing more than designing on paper, and you'll find it becomes ever more essential as projects become more complex. The drawing stage is the point at which major decisions are made about constructing and assembling the parts of the project. If there is a more important phase in a woodworking project than making sure you have a good design, I don't know what it is.

What might be new in this section, even if you already draw, is the notion of drawing projects large and small full scale. If you have been reluctant to venture into the world of design, this section will encourage you to do so. This is the way I've designed and drawn a wide variety of projects for many years, and I show the tools and techniques I use.

Once the project is designed and drawn, you must decide on the best method to build the curved parts. Since no one curve-making method works in every application, the Quick Guide summarizes the pros and cons of each technique for ready reference. Ironically, each of the methods has fewer advantages than disadvantages. This is why radius work is so challenging.

Parts Two, Three, and Four discuss the three primary ways to produce curved parts, and break the methods for curving wood into techniques, tooling, and materials. Part Two focuses on milling curves out of solid stock. This requires either moving a machine like a router across the stock or moving the stock across a band saw, router table, or perhaps a shaper. While most shaper operations can be done with a table-mounted router, I typically use a shaper. When a router is straining to its limit to make a cut, it's time for a shaper to take over. Though the shaper is a much larger tool than a router, operations done on it are remarkably similar to those done on a router table.

Part Three of the book covers lamination-bending. When you see a circular handrail of recent vintage, chances are that it's made up of thin strips of wood glued together to form the handrail contour. Even high-quality furniture pieces are manufactured using lamination-bending. (A well known example would be Thomas Moser's famous continuous-arm Windsor chairs.)

Part Four discusses steam-bending. While steam-bending is fun to do, for most projects it simply isn't practical, which you'll find out as soon as you become familiar with the material discussed. Therefore, the main goal of this part of the book is to shorten the learning curve by explaining the theory of what actually happens when steam-bending works—and when it doesn't. The idea is to make problem-solving easier so success can come a bit earlier in the game.

Part Five of the book puts the methods discussed in Parts Two, Three, and Four to work. Sometimes the best way to make a component combines different curve-making methods. Steaming the laminates and bending them prior to gluing is one example. Also included are techniques that fall through the cracks between the more typically used methods, such as kerf-bending and coopering.

Over the last 20 years of building wooden staircases, wooden boats, and furniture, I've learned a lot about working with curves, usually the hard way. Besides sharing what works, I share the mistakes I've made and talk about how I decide which technique to use on what part. Since no discussion of machinery operations is complete without talking about setups, jigs, and fixtures, I cover these subjects when appropriate. From a radius-cutting router jig using nothing more than angle iron and threaded rod, to instructions for building an efficient steam-bending setup for less than twenty dollars, everything I've learned is here. It's my hope that with this book at your side, your forays into the world of curved woodworking will be both satisfying and successful.

A Quick Guide to Curving Wood

USE THIS SECTION TO ACCESS INFORMATION fast. Read the first part to help select a curve-making technique; the rest of the information is organized under the sections where the information appears in the book.

DECIDING HOW TO BUILD A PROJECT

You must decide how to construct components early in a project. All three of the basic curve-making methods have pros and cons.

Milling solid stock

Parts are cut from solid stock. Milling is always my first choice.

Milling parts with a router or shaper is fast and predictable. Read more about this router setup, used to make curved cuts, on pg. 87.

Advantages:

Precise, easily repeatable shapes are possible.

Milling is easy and fast to do once it's set up.

The wood is strong and stable as long as the arc of the curve is less than 90 degrees.

The process is cost-effective if labor and materials are taken together.

Unprocessed lumber may be used, even of a lower grade, since imperfections may be sawn out of the stock.

Disadvantages:

Power-tool operations require great attention to safety.

Since arcs are cut from solid boards, even with careful layout much of the board becomes scrap.

Beyond about one-quarter of the arc of a circle, grain runout weakens the piece.

While joints may be used to continue the arc beyond 90 degrees, they create visual and technical problems.

Cross-grain lines interrupt the smooth flow of a piece. Avoid butting end grain if possible.

Messy but effective, using the right amount of glue ensures good coverage and imparts great strength to a lamination. Read more about the bending and clamping process on pg. 121.

Lamination-bending

Thin strips are bent over a form and glued. Consistent, high-quality results are possible with the right setup and technique.

Advantages:

A bent lamination has high strength over any degree of arc.

The technique offers consistent and repeatable results.

The wood is extremely stable once it's bent.

There's little waste if sliced veneers are used for laminates.

Disadvantages:

Most appropriate glues are toxic; you must protect your skin, eyes, and lungs. (When sanding, glue is released in powder form.)

Glue lines are visible in the finished product.

Gluing is messy; the excess must be continually cleaned up:

- The process is time-consuming.
- Glue must set up for an extended period.
- Sanding is needed after the glue sets.

The process is wasteful of wood:

- Much wood goes up the dust chute when sawn into strips.
- If veneers are used, they add expense.

On the left is a steam-bent piece. On the right is the chunk of wood that would be needed if you were to saw out the same piece (pg. 58).

Steam-bending

Solid stock is heated and bent over a form. This is a fast way to make certain parts.

Advantages:

It's possible to construct almost any degree of arc without compromising strength.

Steaming eliminates glue lines.

The process is fast, once the equipment is set up.

Steaming is the best and only way to reproduce parts authentically for antique pieces.

Steaming allows an efficient and cost-effective use of material, and results in little waste.

Disadvantages:

Working with steam and heat require extreme caution.

It's difficult to get repeatable bends; different woods react differently and some are impossible to bend.

Steaming requires lots of initial setup time and some expense.

It's difficult to control variables.

Bends may be unstable over time unless they are mechanically held in place.

Sanding and milling are necessary after bending, in part because steaming damages the wood.

Green lumber may check after bending, as soon as the piece cools and dries.

Kiln-dried lumber has special requirements:

- Extra care must be used to avoid grain runout.
- Heating may take longer.
- A compression strap is usually required.

PART ONE: DESIGN, PLANNING, DRAWING 29

Full-scale drawings save time and ensure accuracy.

Draw a project in three views when possible.

There are many different types of curves: simple radius, freeform, complex (helix), and sculpted. All drawn curves must be fair.

Some factors that make a curve harder to form:

- The radius gets smaller.
- The thickness increases.
- The degree of arc increases.
- Complexity increases.

You can measure right off a full-size drawing or use it to cut a pattern from which to construct a plywood template. The leg shown above curves on two sides and is straight on two sides. Learn more about this project on pg. 48.

PART TWO: MILLING PARTS FROM SOLID LUMBER 55

Router jigs and fixtures make the work safe, fast, and accurate.

Rules of thumb for milling solid lumber

Test the setup on scrap pieces of the real material.

Make extra parts. Since curved parts require many steps, if a mistake is made, it will take a lot of backtracking to make another.

Do all milling at one time when possible.

Light, multiple cuts make work safer, more accurate, and leave a better surface finish.

Band-saw rules of thumb

Bigger, more powerful saws are better.

Serious cutting can be done only with a well-tuned saw.

Blade guides are essential.

Stock guides are often deficient, allowing the blade to wander.

The narrower the blade, the tighter radius it will cut.

Even high-quality, tuned saws may not cut parallel to the fence: This fact of life takes getting used to. Once understood, it makes setting up a saw so much easier.

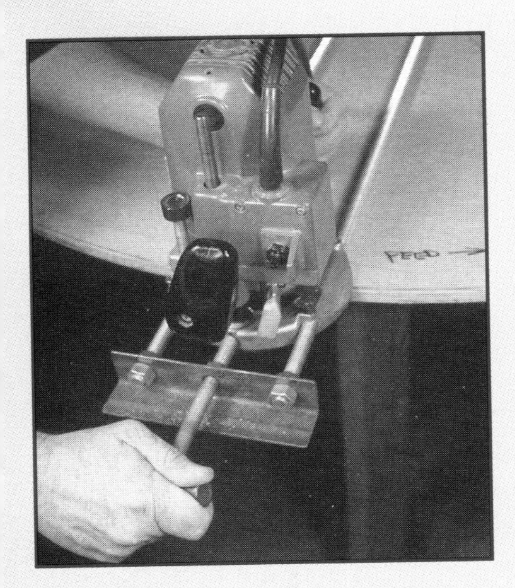

This circle-cutting jig allows a straight router bit to enter the cut from the side, not plunge in from the top. Read all about it on pg. 68.

This jig holds the piece so it can be run by the cutter safely. Arrows prevent feeding in the wrong direction (pg. 78).

Router rules of thumb

Buy the best machine you can afford.

More features equal increased flexibility, which will facilitate future problem-solving.

Large-shank bits perform better than bits with smaller shanks.

A separate collet for different-sized shanks works better than the spacer that some routers use.

Unguided freehand cutting invites problems.

Make light, multiple cuts rather than a single heavy pass.

Cutting curves with a portable router

Cut full circles and circle segments with the router pivoting on a jig. Fasten the work securely to the bench.

Cutting curves with a stationary router or shaper

Setup is sometimes more time-consuming than the actual run.

The piece is run past the cutter as if it were the rim of a wheel with its spokes attached to a pivot point at the hub of the wheel.

The pivot point must be securely attached to the table.

The spokes of the setup will hit the cutter if the piece is pushed too far. The simple solution is to set stops that limit the travel of the spokes, to make sure that they stop well clear of the cutter.

The cutter will need a cutout at the beginning and end of the cut in which it may spin freely.

The machine is started with the cutter free of the piece in its cutout and the spoke against its stop.

The cut is made until the spoke hits the far stop when the cut is finished. Then the machine is shut off while the cutter spins freely in its other cutout.

The inside, or smaller, radius is hardest to set up.

Pattern or flush-cutting

This technique allows the duplication of a pattern's exact shape again and again.

A bearing on the router bit rides on a jig.

The jig prevents the cutter from damaging good wood.

Jigs for holding parts

When using a bearing on a cutter, incorporate both on- and off-ramps for safety.

The shape of the jig or pattern will determine the shape of the part.

Use good material for jig building:

Baltic birch plywood is perfect.

Provide stout handholds.

Make jigs stronger than necessary.

Nice jigs and setups take time, but planning and building jigs can be great fun as well as adding safety and precision to woodworking.

Heavy oak reinforced with plywood makes this jig rigid and super strong. It's used to make curved handrail fittings. *Learn more about it on pg. 79.*

Flush-cutting with a guide batten

A batten may be used to guide a curved cut using a router.

The router uses a shank-mounted bearing.

The batten is brad-nailed to the workpiece along the line:

Check the batten for fairness prior to the cut.

The workpiece is band-sawn close to the line.

The router rides on top of a pair of battens with the one on the line serving as the guide.

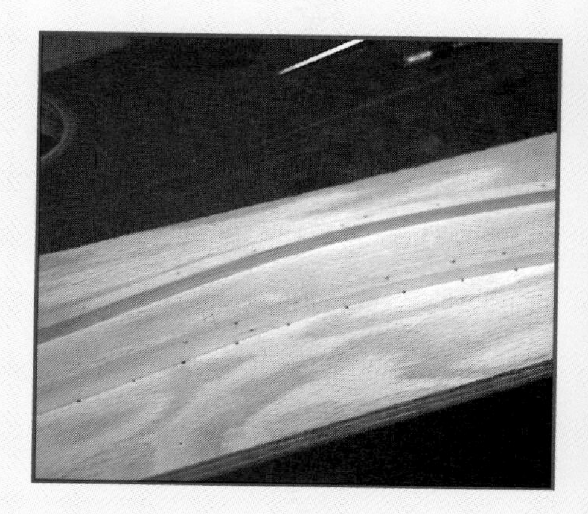

These two battens have different functions. The one on the line is the guide, the other is for the router to ride upon. When used together, they make short work of curved cuts. *Find out everything about this technique on pg. 87.*

Safety with machine operations

Draw large arrows on the table to remind you which way the cutter spins.

Avoid distraction—disconnect the phone, lock the door.

Keep fingers in your sight and out of the line of cut.

Keep both hands on portable tools.

Small clamps often loosen from vibration—don't use them for machine operations.

Use solid, high-quality, sharp blades, bits, and cutters.

Recheck bolts and collets for tightness.

Don't use big cutters in small machines.

Move deliberately, stay focused.

If something doesn't feel or sound right, stop.

Keep tools unplugged unless you're about to turn them on.

Make sure you have good light and footing.

Make off-switches easily accessible, foot-operated if possible.

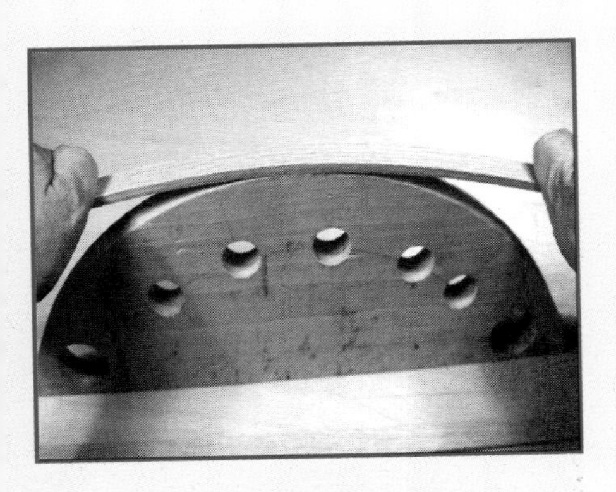

The thinner the laminates are, the better. This laminate isn't going anywhere because it's way too thick to bend around the form. You can learn more about laminate thickness on pg. 100.

PART THREE: LAMINATION-BENDING 95

Make the form as perfect as possible.

Color-match the laminates.

Make a dry run.

Glue all the laminations at one time.

Work quickly when gluing.

You can't have too many clamps.

Never use white or yellow glue.

Determining the maximum thickness of the laminations

The smaller the radius, the thinner the layers should be.

The thinner the layers, the less springback you'll have.

Each layer must be thin enough to bend easily around the radius.

If milling laminates, keep planing the strip thinner and thinner until it bends easily. Then mill enough strips to make up the required thickness.

Using veneers for laminations

This is usually the best choice for bending—it offers the best flexibility and overall quality of bend.

A heated veneer will hold a bent shape even before it's glued.

Flitch-cut veneer lets you preserve the look of solid wood:

A flitch is a bundle of veneers cut from a single log. The veneers are stacked atop one another to re-create the original shape, grain, and color of the log.

Milling laminations on the table saw

Keep track of pieces as they're cut, so they go back together with matching grain.

Handle the veneers carefully to keep the edges from being damaged.

A power feeder on the table saw makes safe work of cutting laminations.

These veneers are thin enough to bend to almost any radius. Keeping them in the same order that they came off the log can result in some beautiful grain matches (pg. 113).

Bending plywood

Plywood laminations can be used to build up a curved shape of almost any size.

Plywood veneer can be used to wrap a curve:

- Thin plywood will easily bend around a large radius.

Two ways to bend around a smaller radius are:

- Belt-sand the back longitudinal layer to expose the vertical layer.
  - This makes the plywood much more flexible.
- Use plywood made especially for bending.

Making a bending form

The curvature must be as close to perfect as possible, with no lumps or hollow spots.

The form must allow for springback.

Rules of thumb for predicting springback

The thinner the laminations, the less springback.

More layers mean less springback.

The larger the radius, the less springback.

The more flexible the material, the less springback.

- Maple is stiffer than poplar, for example.
- Wood is more flexible warm than cold.

Thin layers, and lots of them, mean that this lamination has practically no springback (pg. 104).

Glues and gluing

Harder, rigid glues keep laminations from slipping.

Plastic resin (urea formaldehyde), boatbuilding epoxies, and resorcinol are suitable.

A small roller makes the process of spreading glue onto the laminations speedy indeed. Read more tips on pg. 119.

Soft, less rigid glues are unsuitable because they may allow pieces to slide on one another, or creep, allowing the bend to straighten. Examples are PVA (polyvinyl acetate) glues, including common white and yellow woodworkers' glues.

Gluing tips

Make sure the glue is fresh and work as fast as possible..

Handle adhesives with care. Read and follow the directions and cautions on the container.

Clamping pressure, temperature, open time, pot life, and shelf life all vary according to the type of glue and its manufacturer.

Always do a dry run first.

Apply glue to both surfaces to ensure even absorption.

If the glue skins over before pieces come together, start the process over or risk delamination.

Be mindful of temperature. Almost all glues are affected by it.

Shop-built clamps

PVC sections:

Use sections of PVC, cutting out a segment for clamping.

Clamping pressure is affected by pipe diameter, wall thickness, width of clamp, and opening cutout.

PVC clamps are fast and inexpensive to make in large quantities.

Clamping pressure is low, but these clamps are perfect for epoxy.

These clamps are just sections of PVC pipe sliced off the whole and treated to a cutout. Find out more on pg. 111.

Rubber strips:

Cut up inner tubes. Use the strips to wrap the bundles of laminates tightly. Inner tubes are often available free at tire shops.

The strips are fast to apply, but offer low clamping pressure.

Strapping tape:

Contains lengthwise fibers, has high tensile strength. Use it to tightly wrap a bundle of veneers.

Tape is surprisingly effective for clamping.

Commercial clamps

Bar clamps:

Initially expensive, but after many uses the cost of bar clamps per use becomes surprisingly low.

Many sizes and configurations of clamps are available.

Spring clamps:

Like big clothespins, these clamps are fast to apply, but offer little clamping pressure.

These clamps are suitable for epoxy since epoxy requires less clamp pressure than most glues.

Hand screws:

Slow to use for gluing.

C-clamps:

These are slow to apply but provide high pressure.

PART FOUR: STEAM-BENDING 131

Rules of thumb for steam-bending

The process can seem impossible or simple, depending on how it happens to be going.

Results should improve as you continue to become familiar with bending principles.

Bends most commonly fail because fibers pull apart.

A bend is easily straightened a bit, but it cannot be bent farther.

Always cut bending blanks somewhat larger than net size to allow for sanding and milling.

Wood is damaged somewhat by steam-bending, so removing the outer layer is beneficial.

Make your setup bigger and stronger than seems necessary.

Your setup is at the heart of a successful steam-bending experience. Explore all the options on pg. 153.

The bending form must allow for some overbend to compensate for the inevitable springback.

If the bend is severe, using a compression strap is essential.

No matter what material you're trying to bend, set up using oak. If you can't bend oak successfully, your setup isn't good enough to try other species.

Safety concerns

Live steam is invisible and very dangerous.

Boiling water in large quantities is very dangerous.

Your heat source must be under control at all times.

Wear gloves when loading and unloading the box.

Don't run steam under pressure.

Why do bends fail?

The outside of the curve stretches too much and the piece breaks; tension pulls the fibers apart.

Wood is more easily stretched than compressed, so as it bends, it tears itself apart on the outside surface.

There is insufficient heat or compression.

Too much compression on the inside causes the piece to collapse.

A compression strap limits the stretching of the wood and helps minimize failure. This strap is shop-built (pg. 157), and can be used when working alone with help from a cable puller or block and tackle.

Steam-bending variables

Type of wood:

Woods suitable for bending are oak, beech, ash, and hickory.

Woods unsuitable for bending are teak, walnut, mahogany, and most (if not all) softwoods.

Moisture content:

The higher the moisture content:

The more pliable the wood will be.

The better the wood will conduct heat to its interior.

The more it will check as it dries and cools.

- The less stable it will be dimensionally.
- Shrinkage is an ongoing and inevitable problem with green wood.

Kiln-dried lumber works fine for bending. It's dimensionally stable when cool, resulting in little or no shrinkage:

- Its moisture content returns to the ideal level soon after steaming.
- It's slower to heat through.
- It takes more compression.

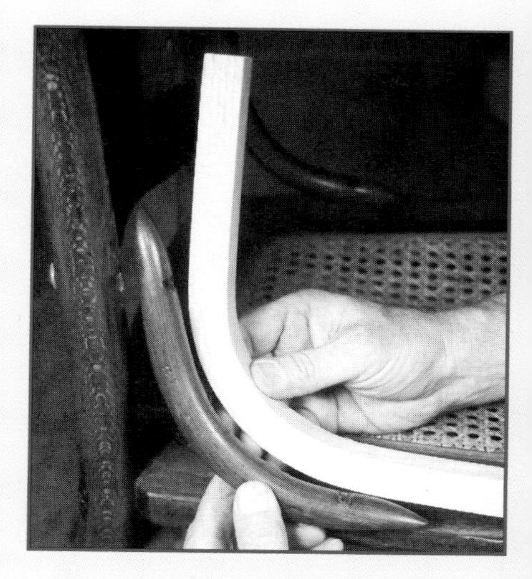

This is a hard bend to steam, thanks to a 1-inch thick 90-degree arc (pg. 135).

An ordinary meat thermometer inserted into a hole in the steam box gives you the control you need over steam-box temperature. Bone up on temperature on pg. 138.

Material-related variables:

- Location of the piece in the tree:
  - The part of the tree closest to the ground is the toughest. It sustained the most strain as the tree grew.
- Grain direction:
  - Where the grain angles off the edge of the piece, the bend will usually fail unless a compression strap is used.
- Flaws in the wood.
- Milling after the piece has cooled:
  - This may cause the bend to relax, another reason to allow for ample overbend.

Severity of the bend:

Bending difficulty increases as:

- The radius gets smaller.
- The degree of arc increases.
- The thickness and width of the stock increase.

Temperature and circulation:

- The steambox should have a temperature of at least 200 degrees at all times.
- The box should have holes for checking temperature.
- Steam should be able to circulate around the pieces inside the box.

Screw-tensioning devices on compression straps allow you to control the bend while you work. This control can spell the difference between the success and failure of a bending project. Get the whole story on pg. 145.

Length of time in the box:

Allow between one and three hours per inch of thickness. Much more than that and the wood will dry out like overdone poultry.

Start watching the clock when the box reaches 200 degrees with the pieces inside.

Compression:

The amount of compression applied with the strap should be the minimum needed to bend the piece.

If the strap has a screw device for tensioning, the pressure may be gradually released during bending:

This adds control to the process.

Leaving the blank on the form:

The blank may usually be taken off the bending form right away.

Clamp the ends to a stretcher or cooling form.

If left on the cooling form to cool and dry, the blank will have less springback.

Controlling variables

It's impossible to control everything.

Keep a log to time the duration of heating for each piece.

Practice moving quickly when bending.

Even though bending is done slowly and steadily, getting the piece from the box to the bending form must be done as quickly as possible.

Bending setup—the boiler

The more BTUs the better. The boiler must supply enough steam to the box to keep it hot.

The bigger the steam box, the bigger the boiler must be.

Electric boiler:

Easy to build with an old deep fryer or tea kettle to boil the water.

Gas boiler:

In good weather, a gas BBQ is a good heat source:

The water container sits on the grill area.

Hoses connect the boiler to the box.

Bending setup—the water container

It should be heavy metal.

Water will need to be added during the process in some way:

The level of the water will need to be measured.

A sight gauge is perfect for this.

A hose or hoses must take the steam from boiler to box:

Hoses should be insulated.

Plastic pipe makes an okay steam box as long as you support it. Otherwise, it will sag in the middle as the temperature inside rises. Pg. 153 gives the details.

Bending setup—the steambox

Use any material that will withstand heat and moisture.

Plastic pipe will soften and sag when heated.

Avoid steel parts (steel can leave black marks on oak).

Make the box just big enough for the wood being steamed.

Remember, the bigger the box, the bigger the boiler must be.

Make a door that is easily opened and quickly closed.

Build legs, so the box doesn't sit on the ground.

Add shelves (dowels are good) so steam is able to circulate around the wood.

Drill holes for steam circulation and thermometer readings.

Plan a drain for condensation to direct runoff.

Insulate around the box and hoses to lessen the fuel required to heat the box.

Bending setup—the bending form

The form must be as strong and smooth as possible.

Sand the bending surface to get rid of lumps and hollows.

The form must provide for clamping in a hurry.

Design the form to allow for overbend (20 to 30%), to compensate for springback. Too much overbend is better than too little.

Glue and nail the form into a solid unit.

If the form is small, design it so it can be held securely to the benchtop.

Bending setup—the compression strap

There's no such thing as a strap that's too strong.

It must withstand thousands of PSI of end pressure.

Avoid steel unless it's galvanized or insulated.

Add handles and backer blocks for strength and leverage.

The ability to adjust compression is desirable:

A screw device is best, but wedges will work.

Hot-pipe bending

This process uses dry heat, not moist, but otherwise applies the same theories as steam-bending.

Hot-pipe bending is typically used to make musical-instrument sides.

Exercise caution, since the heat source is a propane torch, which can easily set clothing afire.

This lovely stair was built using a combination of different methods. Read about it on pg. 166.

PART FIVE: COMBINING TECHNIQUES & OTHER METHODS 165

Steam-bend and glue laminations, then rebend

Steam laminations briefly, bend the bundle around the form, let cool.

Let the bundle dry for at least 24 hours.

Apply glue to both sides of each layer, glue and clamp normally.

Steam-bend solid lumber, then mill

Cut the blank oversize, bend, let stabilize, then cut to net shape.

Bend, using a compression strap.

Let the bent blank stabilize for several days.

Lay out the pattern and cut as you would normally mill solid lumber.

Kerf-bending rules of thumb

The closer the kerfs, the easier the wood will bend and the less cross-grain fracturing there will be, but the piece will be weaker.

The deeper the kerfs, the less fracturing there will be, but the weaker the piece will be.

Regardless of radius, the kerf should be no more than about 1/8 inch or less from the outer surface of the bend.

Veneering a kerfed substrate is easy to do and looks good.

When coopering, a greater number of narrower staves is better than a lesser number of wider staves because it makes it easier to shape the assembly. Find out why on pg. 175.

Veneering a milled inner core

Cut just about any shape you like out of a solid panel.

Make multiples.

Stack and glue.

Sand the shape smooth and veneer once the glue cures.

Wrapping a plywood tube with veneer

Plywood tubes and other curved shapes are commercially available in many sizes.

Veneer is held to the plywood with contact cement.

Building a coopered panel

Coopered panels make beautiful cabinet doors.

A number of staves form the panel.

Narrow staves are better to use than wide staves:

- They're easier to round to a curve.
- Less material needs to be taken off to fair the curve.

This curved door has a coopered panel. The curved rails and straight stiles are milled from solid stock. A piece like this can add zing to any project. Read about how to do it on pg. 186.

Building a curved cabinet door

The technique of coopering is put to good use.

A full-scale drawing is key.

Milling is usually the best choice to construct the rails.

PART ONE

Design, Planning, Drawing

Design, Planning, Drawing

SOME WOODWORKERS are perfectly capable of dreaming up a project, making a sketch on the back of an envelope, and, in what seems like just a few short days of sawing and hammering, producing a wondrous piece of original furniture exactly as initially envisioned. For the rest of us mortals, however, this notion is not only unrealistic, it can be the source of much frustration as we find ourselves painted into various corners for the lack of a plan.

If there is a more important phase in a woodworking project than design, what might it be? What if the apron on the table you just finished is too low for the armchairs to fit under? This is the kind of mistake that drives designers and woodworkers crazy and sometimes to the poorhouse as well. Without sound design and good drawings, you're relying on luck to keep you from making serious errors—and luck never holds forever.

Here's an example. A stairbuilder I know was called out to look at a staircase design in an expensive new home under construction. The home was in the advanced framing stage, with nearly everything but the staircase in question finished. According to the plans, the stair had an intermediate landing with a powder room located right beneath. A great use of space, the architect thought, until a small problem surfaced.

After just a few minutes of layout, my stairbuilder friend found that the landing was about 6 feet above the first floor. This unfortunately meant that the ceiling height in the powder room would have been less than 5-1/2 feet. Whoops.

The rather tall, normally tolerant client was not amused. About the only good news was that the problem was discovered early enough in the project to be solved. And how do you think it was fixed? The stairbuilder sketched some fairly simple working drawings of a new design for the staircase as well as a new location for the powder room. As it turned out, the low-ceilinged space below the stairs made a perfect closet for holiday decorations. With relatively little time and effort, everyone but the very red-faced architect was happy again.

WHY SPEND TIME DESIGNING AND PLANNING?

Obviously, the purpose of designing and planning a project is to have it come together just like you want it to the first time. Paying attention to the details doesn't mean you'll be guaranteed fail-proof success during every step of a project, just that you'll have more control over the work from beginning to end.

The first task in design is to determine the purpose of the project you wish to build. Before starting on a chair, for example, you probably would want to address whether your end goal is comfort or art. Answers to many questions will flow from your decision. Should the design match a certain style or be completely original? Are arms needed? Is weight important? Should woods and finishes match or contrast with other furniture in the room? Will the piece fit the space it is supposed to occupy? Will it fit through the door?

You design your project around its purpose and plan its construction before beginning to build. This is where drawing comes in. When you first see something in your mind's eye, the tendency is to view it somewhat optimistically. Everything fits perfectly. There are never any construction problems. A corner here or a joint there might be a little bit fuzzy, but the whole is just right. This vision is often what drives a project.

However, once you take into account real proportions, things sometimes don't work as well. Without a drawing, it's difficult to decide if a design is pleasing to the eye. If you are working for a client, it will be exceptionally difficult to convey your ideas without something on paper. Sometimes the drawings will lead to a prototype—an experimental construction that tests whether the joinery and the design will work like they're supposed to. Perhaps you'll be using outside help to make some of the components, in which case good working drawings can illustrate precisely what is required even to people who don't speak the same language. In addition, drawing can give you increased confidence about the project as a whole. There is no more important job in woodworking than making sure you have a workable design, and no more important design tools than the pencil and eraser.

By learning to draw, you can develop your own designs from scratch or modify existing designs. Let's say that you want to build something similar to but slightly different from an existing design. For example, you might want to modify a chair into a settee, a table into a desk. If you try to do this without drawing, you can throw off the entire project. "But it's only an inch," you say. Well, just try adding or subtracting that inch from any set of plans. It will change the cut list, distort the proportions, and affect the project in more ways than you can count. While you may think you are saving time by skipping the drawing step, in fact you'll be doing exactly the opposite.

CONTROLLING CHANGES

Any modifications that you (or a client) desire can be made far more easily with an eraser than a saw. A series of drawings also forms a paper trail, which can be a valuable reference as work proceeds. If a client decides that the chair must have arms after all, it's straightforward to draw them into a drawing with the implication that changes in the drawing mean changes in the fee. More importantly perhaps, the addition of an arm to a chair may necessitate complete redesign. Again, this is more easily done on paper than in the mind while the saw is running. ■

DRAWING FOR SAFETY

Making changes can actually be a safety hazard. Have you ever stood at the table saw and found yourself wondering if the piece should actually be ripped at 4 inches, not the 3-1/2 inches at which the fence is set? If you have your drawing handy, it's simple to shut off the saw, walk over to the drawing, make a measurement (easily done with a full-scale drawing), and return to ripping the material with new confidence—the perfect frame of mind in which to safely operate a saw. ■

WORKING DRAWINGS AND THE CASE FOR FULL SCALE

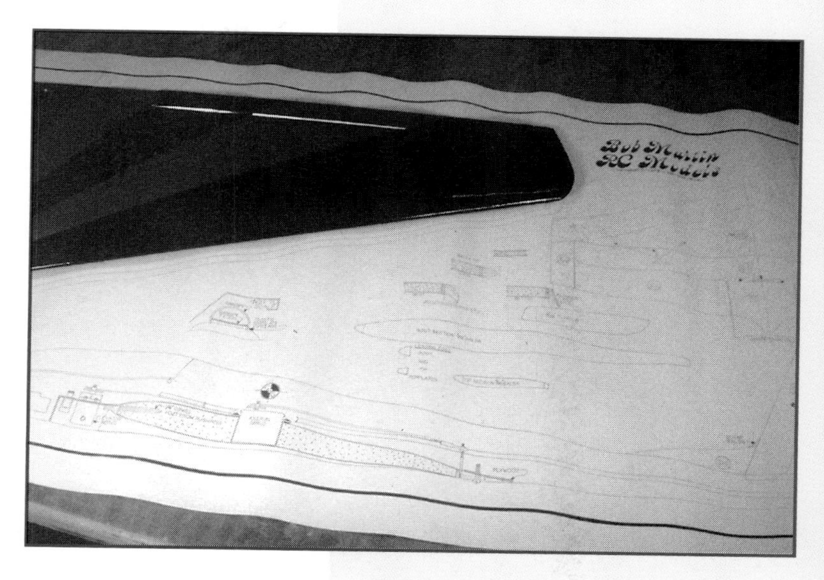

This is the way most model airplanes are constructed: using a full-size plan. Boats and ships, even very large ones, have been drawn full size, or "lofted," for centuries.

A working drawing is the drawing from which you actually work. If you have ever built a model airplane using balsa ribs and spars, you probably remember that you build the parts right on top of a full-size drawing. This way each part can be positioned exactly according to the plan. By the time the airplane is finished, the drawing is typically full of pin holes, notes, and glue dribbles. Likewise, by the end of a woodworking project, your working drawings should be wrinkled, full of erasures, notes, and stains from your favorite coffee cup.

Full-scale drawing is an easy process. There are no scales to learn, no dimension lines to draw—and when you're finished, you can measure right off the drawing to size the parts. The drawings are accurate and easy to follow, and there is less potential for error through common mathematical miscalculations or mistakes in converting scales. Obviously, the more complicated your projects get, the more essential drawing becomes as a way to rein in an otherwise mind-boggling array of measurements and parts. Especially when working with curves, you have to make sure that the curved components will mate properly with the other pieces. In the course of drawing a project full scale, you can make decisions, design jigs, and get organized.

Lastly, drawing full scale can be really enjoyable. When the project unfolds on the paper for the first time and actually looks like it will work, it's a great experience.

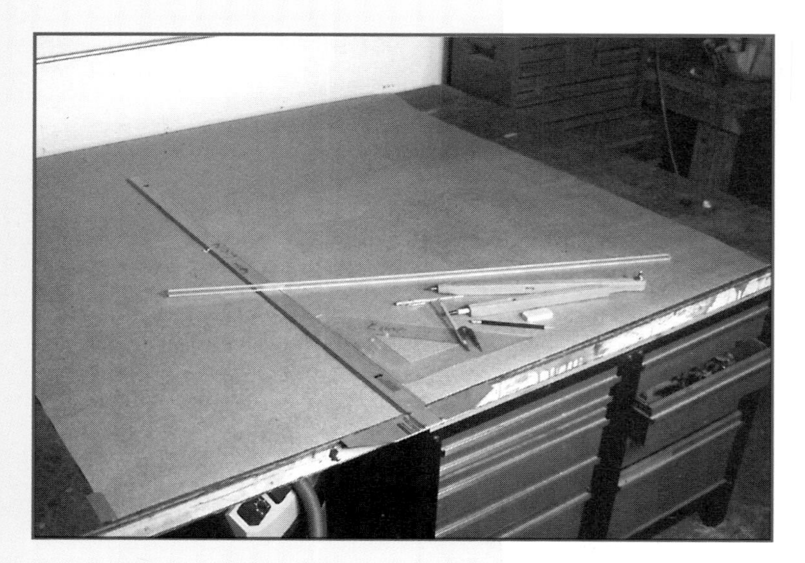

The drawing board is where I do most of my thinking. When in doubt, back to it I go. A few simple tools, the most important of all being the eraser, are all you really need to draw just about any woodworking project.

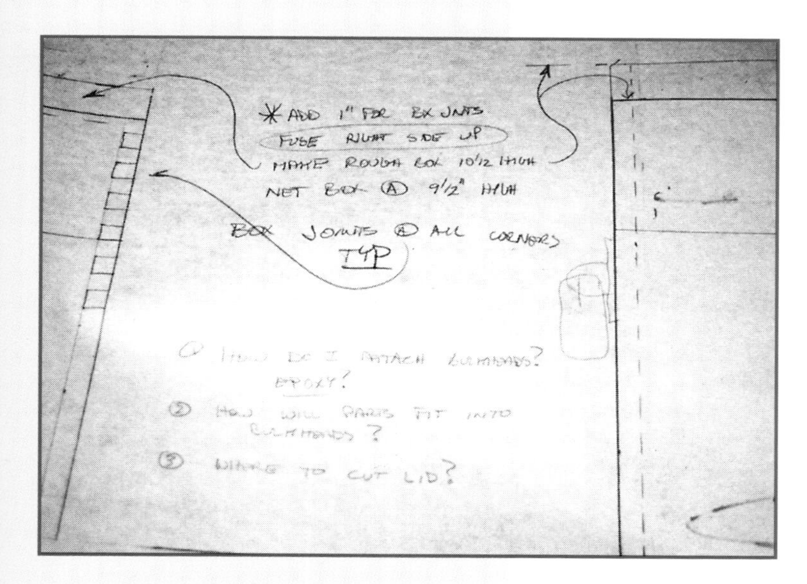

As problems (or "challenges") come up, those I can't solve right away go on a list. I cross them off as I solve them and add more as I discover them. When everything is crossed out—and not before—it's time to get to work.

Using drawings to solve problems

The real value in drawing full scale becomes evident when there is a problem. It's a great comfort to be able to return to your drawing to see what's gone wrong with the project and analyze how to salvage it.

Try as you might, sometimes you won't be able to foresee all the facets of a project, and you will eventually run into some detail that seems impossible. This is the time when real creativity takes place and some other force enters the process. Maybe you'll wake in the night with the germ of an idea. At the drawing board the next day, as if by magic, a solution to your problem will appear on the paper.

Don't try to understand how this works, but it will happen often enough to be almost predictable. So when you reach the point where a project seems like it will never work out, it means only that a breakthrough is probably about to happen. Keep drawing and the solution will present itself. Armed with a straightedge and an eraser, you can almost always sort things out. Just keep drawing in more and more detail and before you know it, you'll be on to other problems, having solved the first batch.

When you discover that something needs to be worked out and you don't have time to devote to the problem right then, make a note directly on the drawing. This way you can go on working without fear of forgetting the obstacle. Plan to set aside an area on your drawing just for this purpose. You can call it your "problems yet to be solved" list.

If you draw over the course of several days, when you come back to the list you'll have an automatic reminder of the hurdles you still need to clear. As you solve problems, erase them from the list or cross them out. When the list is gone, you're ready to go to work.

Determining work sequence during the drawing phase

Sequencing a project is often the most difficult task. Now is the time to think through the project. Your sequence might be as simple as saying that you'll rip the pieces first then crosscut them. But if the project is complicated, the sequence will be much more involved.

If you have a delivery deadline, the sequencing process offers the perfect opportunity to estimate the amount of time to allot for the completion of each task. (To check how realistic your estimates are, make it a habit to record how long something actually takes, then compare that time to what you estimated.)

Using drawings to develop cut lists and tooling

List materials, hardware, and tooling to be purchased right on the drawings. This way you can roll up the drawings and take them to the lumberyard when you go shopping. Although this step is easy to overlook, especially when you're impatient to start sawing wood, getting into the habit of making lists and keeping them close at hand will ultimately help you work more efficiently. As a bonus, if you're organized, you might be able to bring in helpers to cut parts. They'll need to know that you need 46 of these parts this long, this wide, and this thick. That's the essence of a cut list.

You may find that a small change that reflects the tooling or machinery that you already own may save expense. For example, a molding pattern or the size of a component could easily be changed to suit your equipment on paper, but might be difficult and expensive to run if special tooling is required.

THE PROCESS OF FULL-SCALE DRAWING

By now you're probably thinking, "This sounds like I'll need a piece of paper the size of a circus tent!" Not really. Of course you could get carried away with large drawings, but you could also draw even the largest project just a section at a time. Think of the sections as tiles which, when connected to one another, form a large, full-scale drawing.

If you can get used to handling a larger sheet of paper, it will make drawing to full scale easier. Brown craft paper is inexpensive and available in rolls up to six feet wide. The paper used for photography backgrounds (called "seamless paper" by photographers), is of excellent quality if a bit expensive. If you have trouble finding paper that's large enough, a thin piece of plywood or particleboard will work just fine. Try painting it white.

I now use a simple fold-up table to draw on. The paper is rigged so it unrolls right across the table, coming up from below just like a roll of paper towels. But don't think you have to make a major investment to be able to draw full scale: I used a piece of plywood over a couple of sawhorses for years.

Drawing tools

At first it may not much matter what kind of tools you use, but later on you may reach the point where you appreciate good-quality drafting equipment.

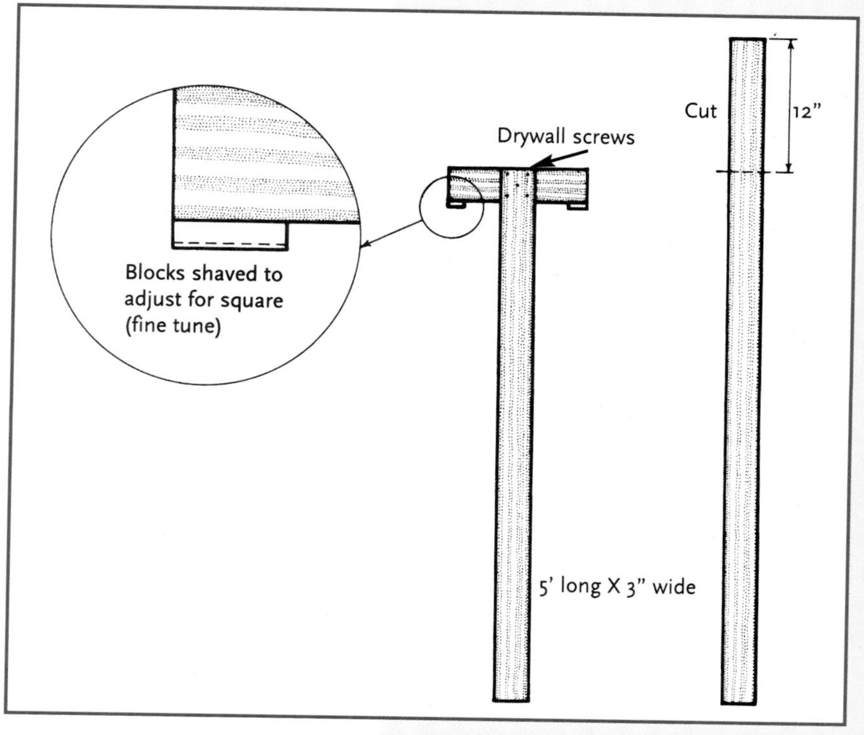

It's easy to make a square that can handle almost any full-size project from a straight length of board.

To begin, you'll need the basics—a pencil, eraser, and paper. (Soft white erasers will allow you to erase lines without damaging the paper.) For curved work, add a compass or trammels, a thin flexible stick or batten, and a ruler or tape measure. A large pair of dividers is useful, as is a set of French curves. A large straightedge is essential. An inexpensive square made to cut drywall can double as a straightedge, or you could even make a square out of boards. If you happen to have a straight, not-too-thick board about 5 feet long by 3 inches wide, cut a 12-inch piece off one end and glue it to the long piece as dead square as you can get it. Use four or five drywall screws to fasten the two pieces together while you adjust it. Work quickly because once the glue sets, it will be too late. If you discover it's a little bit off, glue two small squares of wood onto each end of the short part. Then carefully shave one of the squares a little bit at a time until your new square is perfectly...square.

Drawing basics

Before we tackle curves, we'll draw a simple box to demonstrate how easy the process is. If you already know how to draw, you may of course disregard the following section. But if you haven't ever drawn full scale, you might want to give it a try. Should you be one of those woodworkers who thinks that drawing is something you simply cannot do, please relax, breathe deeply, and try the following exercise. Drawing is not a skill that's inherited at birth. It's a skill that anyone can learn.

The essential three views

Drawing the front, top, and end views of a piece in flat planes makes a complete picture of the project. The front view shows length and height but not width. The top view shows length and width but not height. The end view shows width and height but not length. All three views together show the entire project.

The orientation of the views to one another is fairly standard in construction drawings. The front view goes on the lower left portion of the paper. The top view is positioned directly in line with and above the front view. The end view goes directly to the right of and in line with the front view. By representing a piece in this way, dimensions can be carried from one view to another, making the process fast and accurate.

Try it yourself

1. As shown on pg. 40, draw two construction lines on the paper to represent the bottom and left side of the box. Draw the vertical line an inch in from the left edge of the paper all the way from top to bottom and the horizontal line an inch above the lower edge of the paper all the way across. The lines intersect in the lower left corner of the paper.
2. Let's say the box is 12 inches long. Draw a second vertical construction line 12 inches to the right of the first vertical construction line clear up to the top of the paper.
3. Let's also say that the box is 6 inches high. Draw a horizontal line 6 inches above the first horizontal construction line all the way across the paper. The rectangle formed by the first four construction lines is the front view of the box, a rectangle 6 inches by 12 inches.
4. Now draw the top view. Suppose the box is 8 inches wide. Start by drawing two horizontal lines—the lower one a couple of inches above the front view and the upper one 8 inches above that. You've now created a rectangle measuring 8 inches by 12 inches.
5. To create the end view, draw two vertical lines 8 inches apart

(because the box is 8 inches wide). Draw the first a couple of inches to the right of the front view and the second 8 inches to the right of that. This third rectangle, 6 inches high by 8 inches wide, is the end view. Label each view and darken the three rectangles.

The result—three views quickly drawn to full size—can also contain details such as material thickness, grain orientation, corner joinery, lid operation, and hinge placement.

DRAWING CURVES

Let's attempt a full-scale drawing of a simple corbel design. Start by drawing straight lines to mark the height and width of the corbel.

Suppose the corbel is 12 inches high and 8 inches wide. Draw a horizontal construction line that is 8 inches to the right of the vertical construction line and a distance of 12 inches above the horizontal construction line.

The lower, or thin, part of the corbel will be 2 inches wide, so next we draw a vertical construction line 2 inches out from the left construction line. We will make the top, or thicker, part of the corbel 4 inches tall, so we will draw a horizontal line 4 inches below the top construction line. These numbers are arbitrary. Use whatever proportions please your eye.

We now know the height and width of the corbel, but not the thickness. The top and end views do not provide much useful information at this point, but it is a good exercise to draw them just the same. We'll be working more on the end and top views later.

For now, let's say that the corbel will be 1 inch thick. We can easily complete the three views by drawing a few more construction lines. Leave plenty of room above and to the right of the corbel's front view.

We have drawn all of the straight lines. Now it's time to tackle some curves.

Start with the top of the corbel. Set the compass at a 4-inch radius. Draw a 90-degree circle with the point of the compass on the top

Drawing a simple box

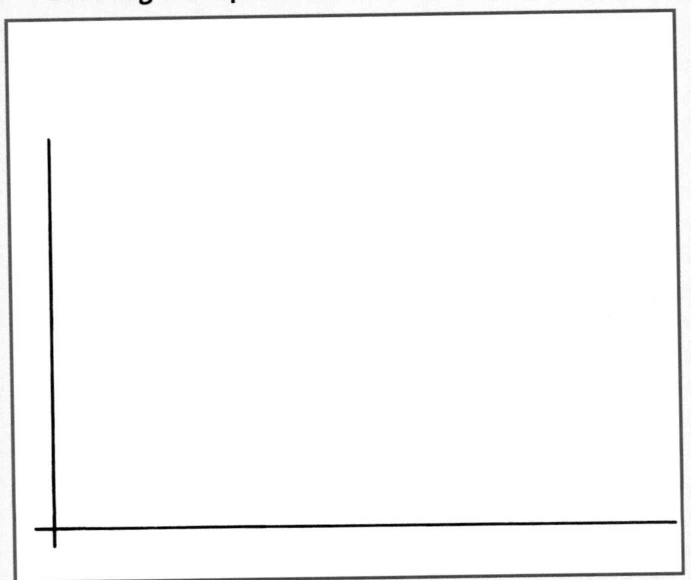

Two construction lines define the bottom and left side of the box.

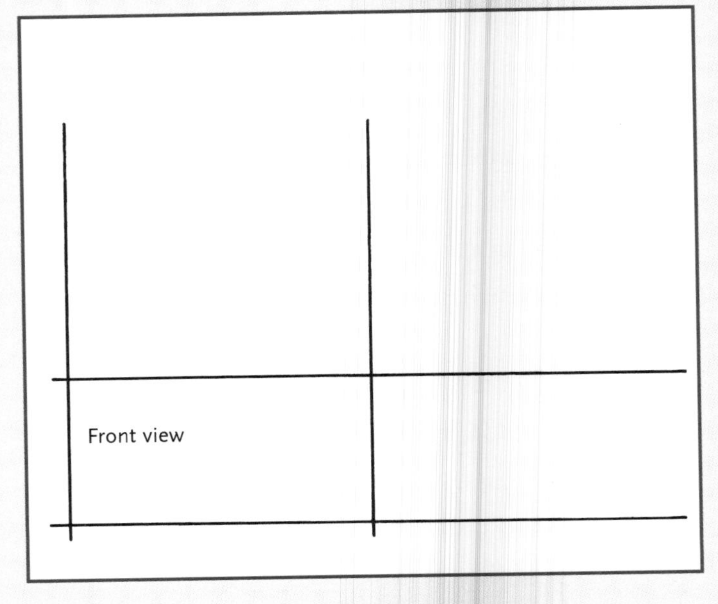

Length and height are indicated by two more construction lines.

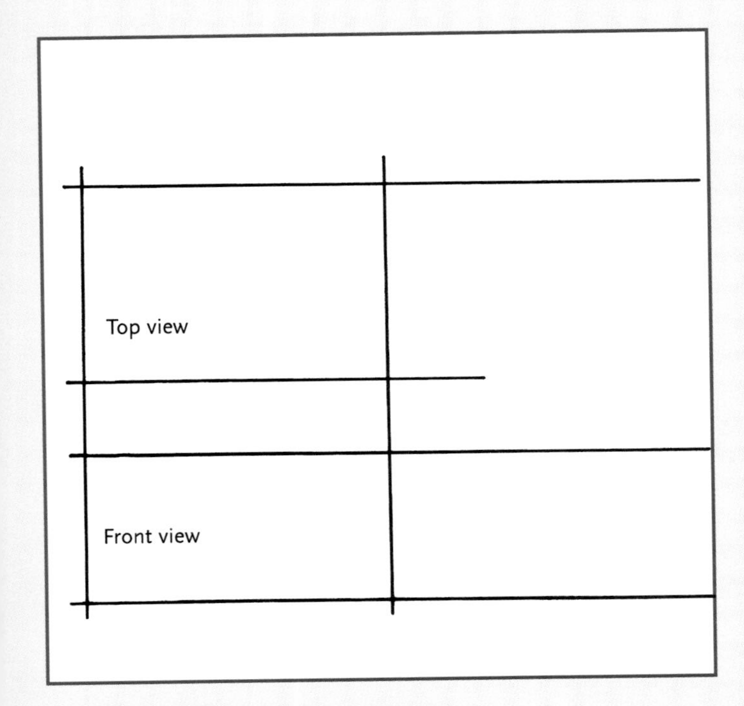

Create the top view by drawing two horizontal lines the width of the box apart.

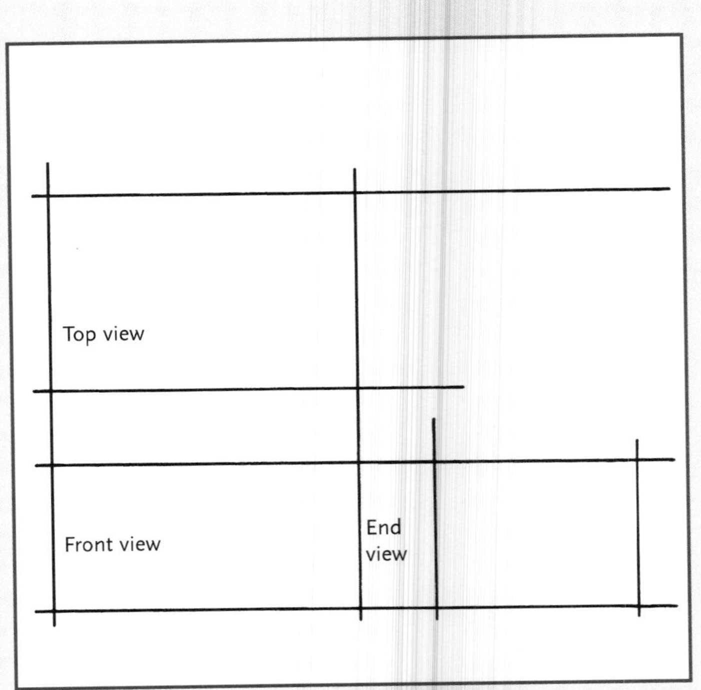

Draw the end view by penciling two vertical lines the width of the box apart.

Drawing a simple corbel

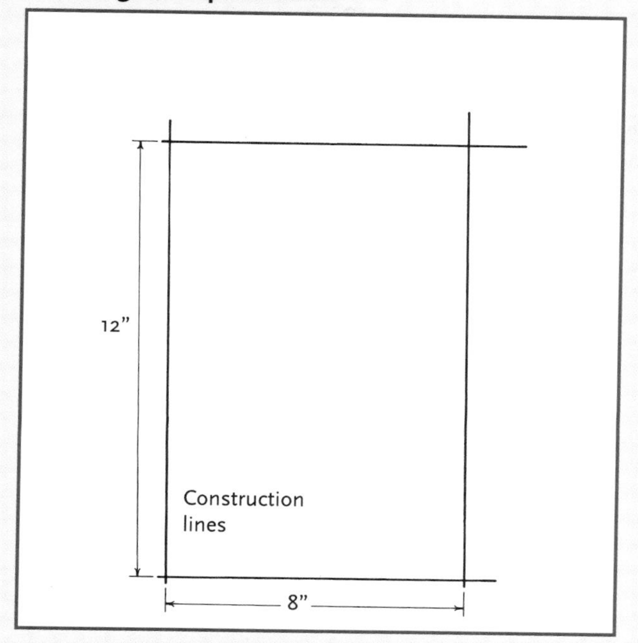

Draw the lines for the width and height of the corbel.

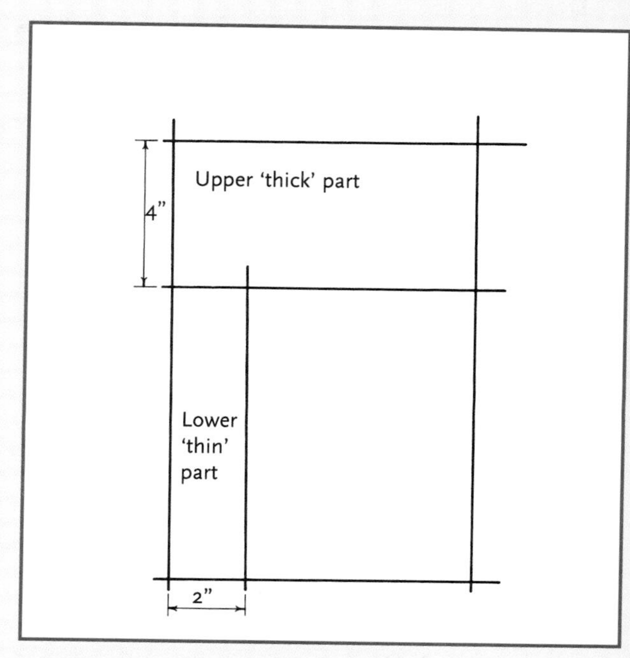

Mark the lower and upper parts of the corbel.

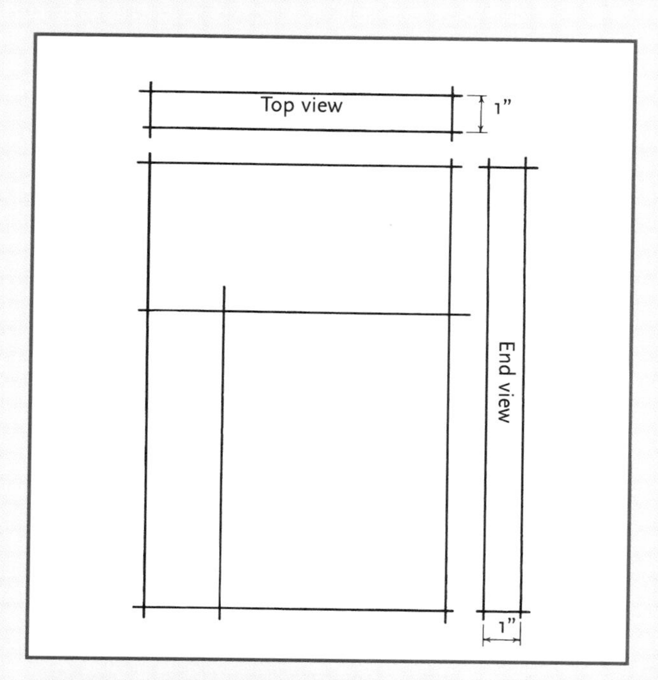

Finish off the last straight lines, showing the corbel's thickness.

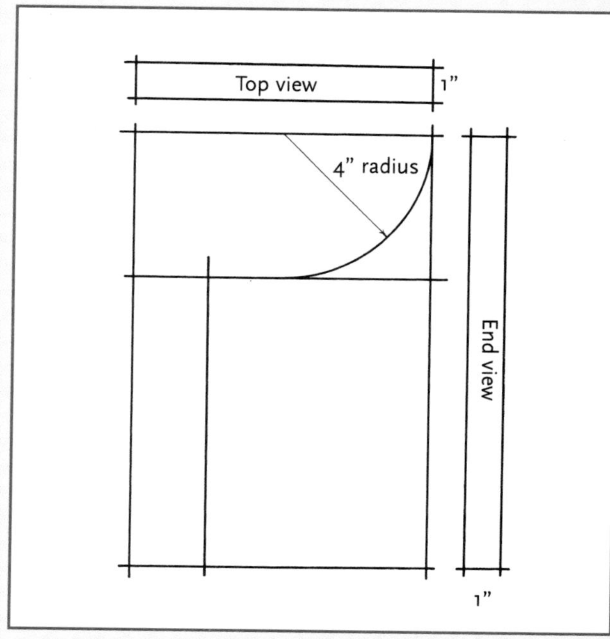

Draw the first curve with a compass.

Variations on the curve

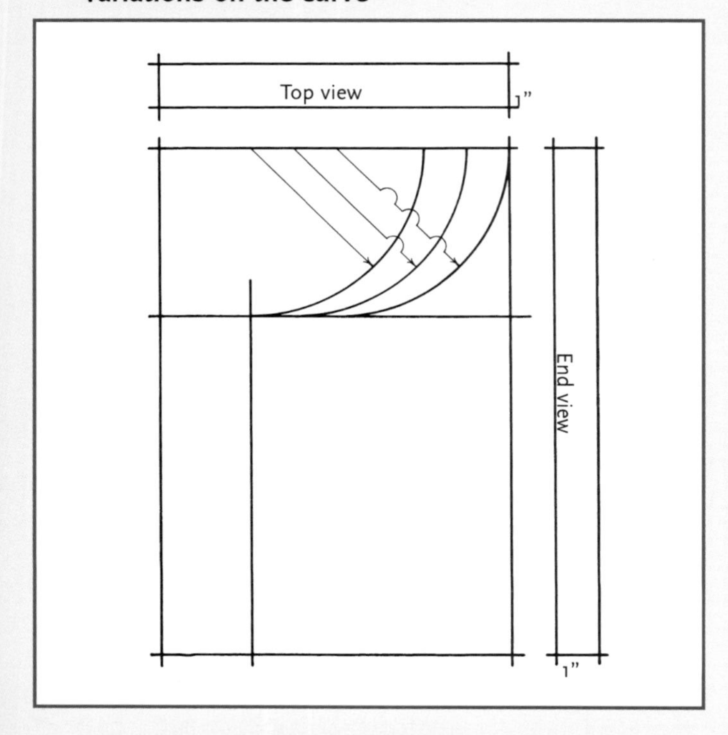

Moving the compass point along the construction line at the top of the corbel varies the corbel's width, but the curve remains pleasing to the eye.

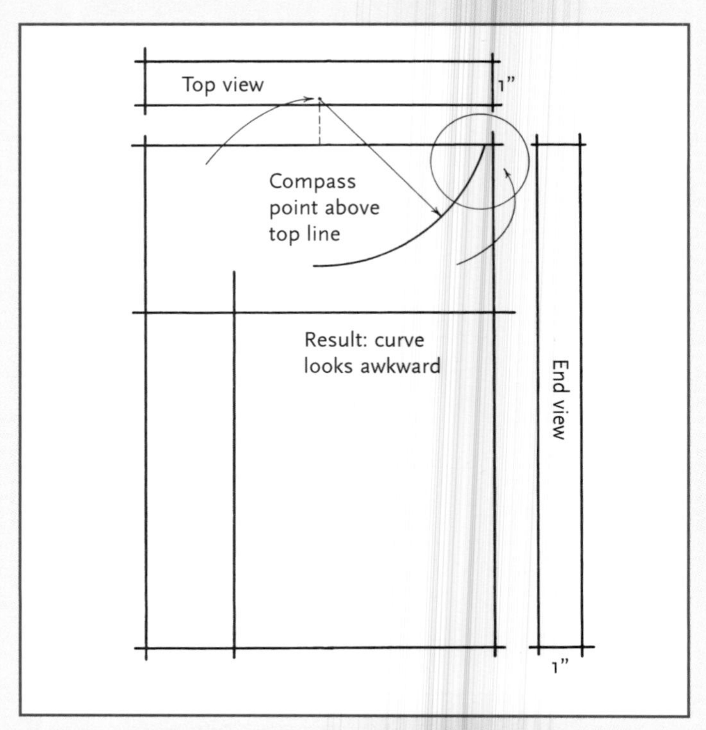

Locating the compass point above the construction line results in an awkward-looking curve.

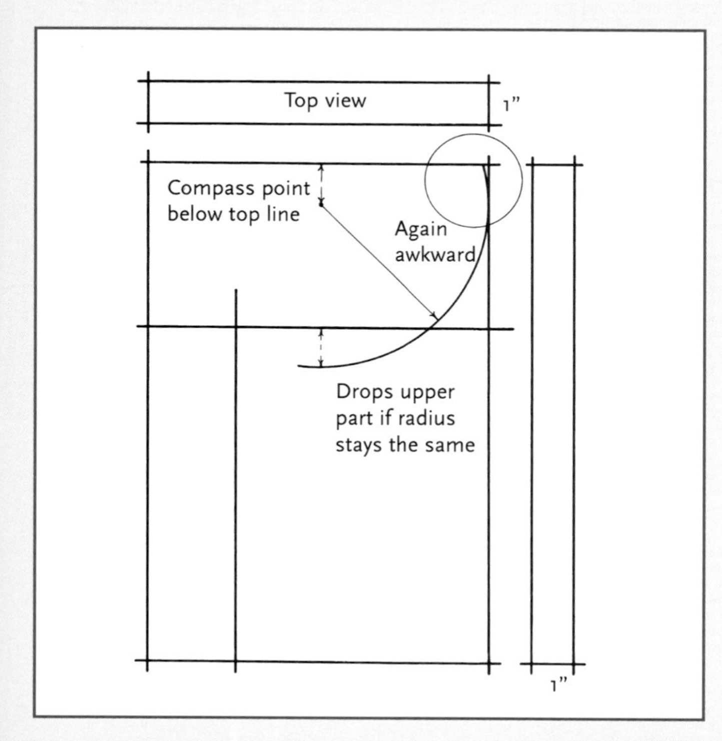

The curve that results from moving the compass point below the construction line has a crevice at the top, but you can cut this off to make a good-looking curve.

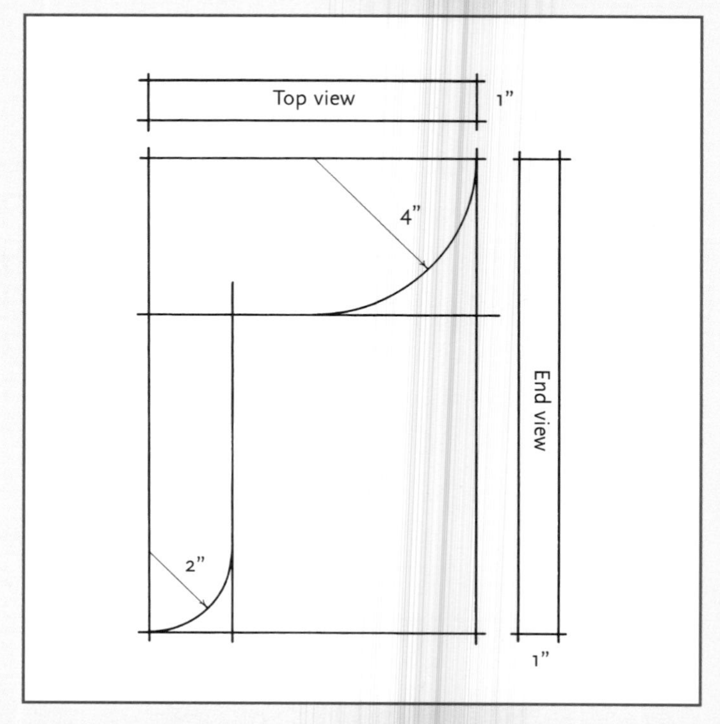

The design takes shape when you draw the curve at the bottom of the corbel.

line and the pencil at the right side construction line. Make a mark on the top line where the point of the compass is. Note that if you put the point of the compass anywhere along the horizontal construction line at the top of the corbel, the resulting curve will begin at the top of the corbel as it should. Changing the location of the compass point along the line simply changes the width of the corbel, but the resulting curve looks correct.

Now let's have some fun deliberately moving the compass point to the wrong spot. Place the point about 1 inch above the horizontal construction line.

See how the curve looks awkward if you start with the point of the compass above the top line? It is too sharp an angle and looks clumsy. If you put the point below the line, the curve hooks back into the top to create a small crevice, but if you draw a tangent vertical line at this point, the corbel once again looks correct. The ideal beginning of the curve is on the top line or below it. If the curve starts with the compass below the line, the top of the corbel remains straight for the same distance.

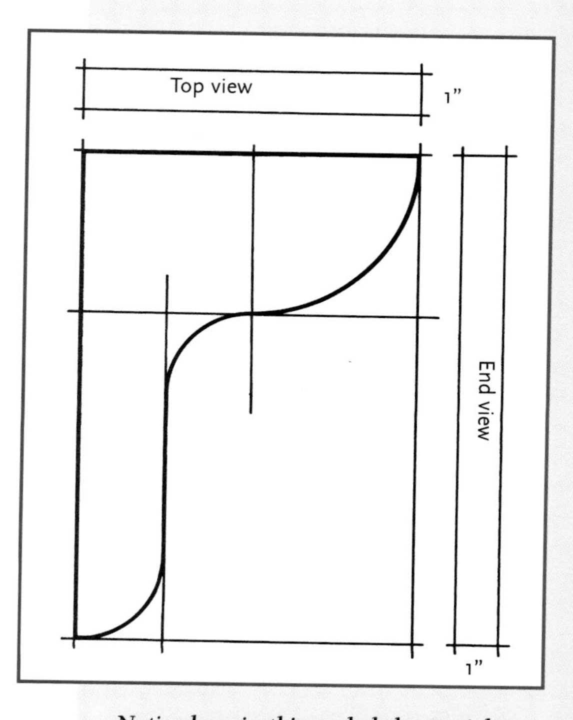

Notice how in this corbel the straight lines blend nicely with the curves. It's all a matter of paying close attention to the location of the center point of the compass.

Much of this sort of thing is subjective. If a curve looks good to you, that's all that matters. If you have changed the lines more than once with this exercise, you are experiencing creativity and becoming aware of how much easier it is to erase a pencil line than it is to re-cut a board.

Now set the compass at 2 inches. Draw a similar line at the bottom of the corbel, starting at the bottom of the corbel and connecting to the vertical construction line.

We have one more circle to draw—the inside curve. Draw a vertical construction line where the compass point intersects the top horizontal line. Note that this line is 2 inches to the right of the narrow part of the corbel. Set the compass to 2 inches. Make a mark 2 inches down from the top circle along the vertical construction line.

Now draw the last curve. This completes the rough outline of the corbel. Leave the construction lines in place and darken the outside perimeter lines so that the outline of the corbel becomes more clear. And note that even though the two upper curves are different sizes, they fit together perfectly.

Variations

The variations on the simple corbel are almost endless. Try using different radius circles, different proportions, and various combinations of straight and curved lines to change the shape of the corbel. This is the creative process at work. When it looks right, it probably is right.

Instead of a flat corbel, let's see what it looks like curving in two planes at once. Our top and end views now come into play once more.

What you just drew was a flat-plane, simple curve shape using only true circles. Let's try one more variation. Say we want to curve the corbel in the other direction. What would that look like?

We have top and end views showing the entire piece 1 inch thick. Let's keep it 1 inch thick at the bottom, narrow, part, but make it 4 inches thick at the top. Start by drawing some straight construction lines.

We want to make a graceful curve to taper our corbel from the 4 inches at the top to the 1 inch at the bottom. The top view shows only a rectangle 4 inches thick and 8 inches long. It is the end view that will show the curved taper we are about to draw.

Draw a construction line down 3 inches from the top of the corbel on the end and front views. For some reason, we do not want a simple curve in this view. (It is as if we are looking for a way to make an easy project hard, but sometimes aesthetics are everything. Such is woodworking.)

Using a French curve, trace a graceful curve in the end view. We can have some fun here since we're only drawing on paper. Once it looks just right, stand back and contemplate how you would build such a thing. Oh, that.

In this example, a construction line drawn 3 inches down from the corbel top on the end and front views is the starting point for creating a transition curve.

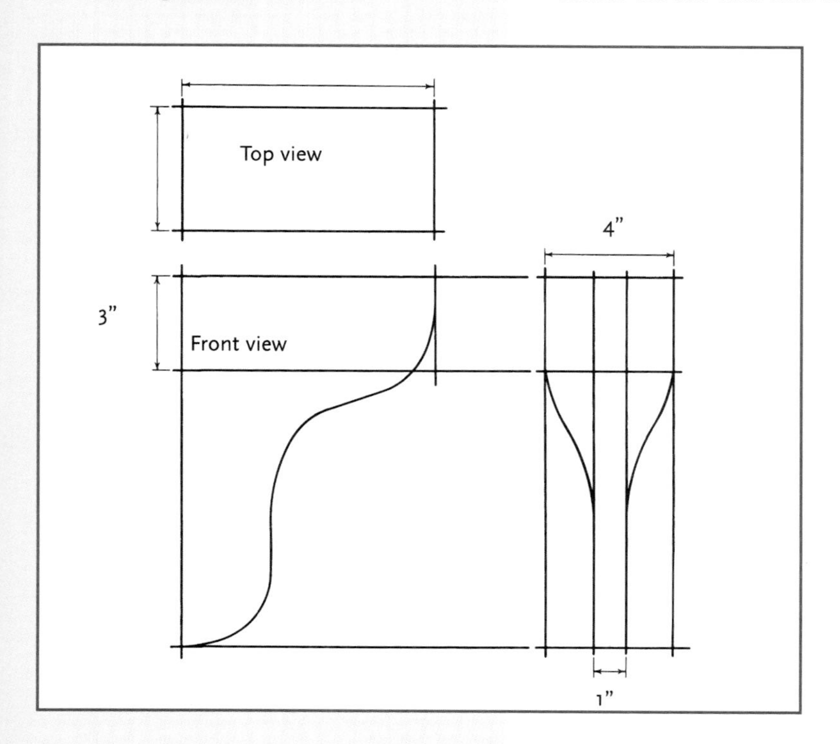

Beyond the basic drawing

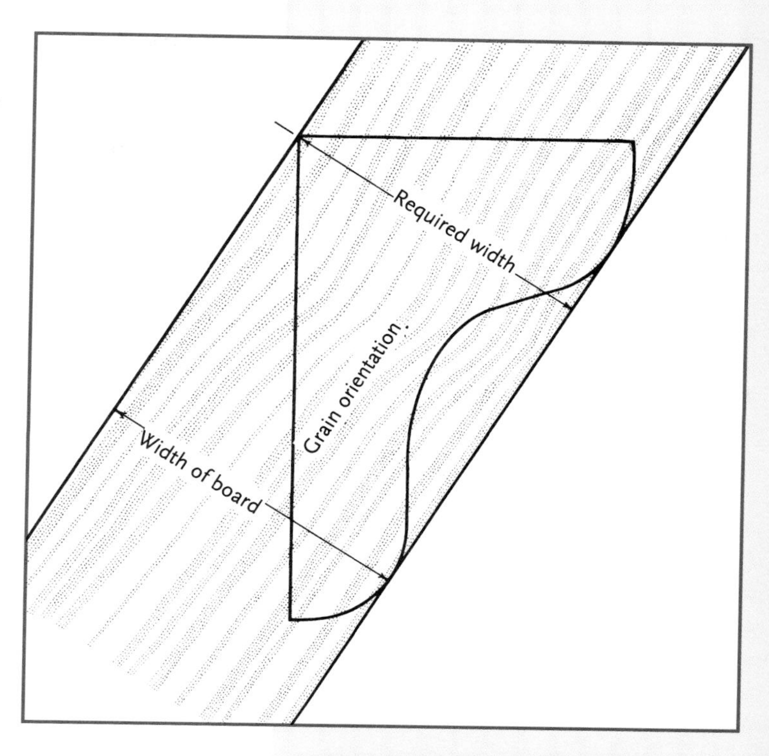

Because the drawing is full size, marking two parallel lines on the corbel will let you determine the required width of the board.

Now that you've done the drawing, visualize a solid block of material 4 inches thick by 8 inches wide by 12 inches long. That's your cut list. Now visualize cutting away with the band saw all but the shape you see on the paper. First you would cut the end view, and since the piece in this direction is 8 inches high, you would need a sizable band saw. Oh… your band saw goes only to 6 inches? Better make a note of that. (Note possible solutions as well—perhaps you have a friend with a bigger band saw you could use?) Then you'll tape the piece back together so you can cut the front profile. Or would it be better to reverse the cut sequence? Jot a note to yourself in the space you set aside for problems yet to be resolved and go on. Next you would do more than a little sanding, and, just like magic, there is your corbel. That's your project, including a drawing, cut list, and sequence list complete with some tooling requirements thrown in.

This is obviously the time to address the question of grain orientation. If the grain is vertical, the projection will be weak due to short grain across its length. The solution is to run the grain diagonally. Not only does this strengthen the corbel, but as you can quickly measure, saves material as well.

On the drawing, create a line that touches the upper and lower projections on the corbel. Now draw a parallel line that touches the upper corner of the corbel. In just a few seconds, you can figure the exact width that the material must be if the grain is to be diagonal. Since the drawing is full size, what you see is what you get. You can measure the width of the board and just like that you'll know for sure how wide it must be.

Don't forget to step back for a moment and assess the big picture. For example, is curving the corbel in two directions worth the extra material and labor required? Would a simple flat corbel do the job? Is a more creative design worth the trouble? Does the look of the piece justify buying a new band saw? These questions are never easy to answer. You must wrestle with the various reasons, rational and otherwise, for choosing one option over another. You get to experience some of the torment the artist in you must suffer.

And that's really all there is to designing a project incorporating both simple and French curves. This is, of course, only the beginning of the process of drawing and design. But it's one that should be simple enough so that those who find drawing difficult will be able to reach a successful conclusion.

UNDERSTANDING DIFFERENT TYPES OF CURVES

Just about any piece that is not straight or angular could be called a curve. There are two- and three-dimensional shapes, radius and freeform curves, simple and complex curves; all are distinctly different and employ different construction methods.

In the process of designing a woodworking project, the complexity of a curve may well drive (or stall) the design. For example, a designer may be able to simplify a project by changing a curve from freeform, for example, to a curve with a constant radius. This may, during the course of construction, save days of labor and a significant amount of material. On the other hand, a more complex curve may make an otherwise plain project so beautiful that just about any effort or expense is justified. These are the sorts of trade-offs with which designers of woodworking projects continually wrestle.

Simple radius curves

The simplest curves are arcs on a flat plane, or constant-radius curves. These curves are simply full circles or segments of a circle. Curves like this can be drawn on paper with a compass. Each point on the curve is the same distance from the center point. The advantage of using simple curved shapes for woodworking projects is that these curves can be produced accurately, quickly, and easily.

Machine-cut simple curves can be made in either of two ways. The first involves moving a portable cutting tool like a router along a constant radius by fastening it to a movable arm attached to a center point. In this case, the workpiece is held stationary. The second method involves the same principle, but the workpiece moves into the cutter of a stationary machine.

Many very complex shapes can be drawn using only constant-radius curves and straight, tangent lines. Circles, or segments of different-radius circles, may be connected to form other curved shapes. Smoothly connecting one curve to another is often simply a matter of drawing some construction lines. This is the principle we explored when drawing the corbel previously.

Freeform three-dimensional curves

On the other end of the spectrum are curves that have no constant curvature. These freeform curves are often found in sculpted furniture. They follow a precise curvature but are anything but simple.

A cabriole leg is an example of this type of curved shape. Often the cuts can be made one plane at a time to rough out the part. First you cut along lines drawn on one face of a square board. Then you rotate the piece 90 degrees for the next cut.

Parts with sculpted curves can usually be cut out at least in profile. Then they can be carved to their final shape.

Helix curves

In between these two extremes are curves like those found on the handrails of circular staircases. These handrails are helix curves, similar to the stripes on a barber pole. A helix is usually a simple curve when viewed from above, but it rises as it curves to form a spiral shape. Most of the time, handrails are bent on a constant incline.

Cutting a curved table leg

The full-size drawing for this curved table leg is used to cut a paper pattern. The shape is then traced on a plywood template. This particular leg is curved on two sides and straight on two sides.

By attaching battens to the plywood, the curve for the template is easy to cut with a band saw and router. The battens should be stiff enough to bend evenly, but flexible enough to bend without breaking.

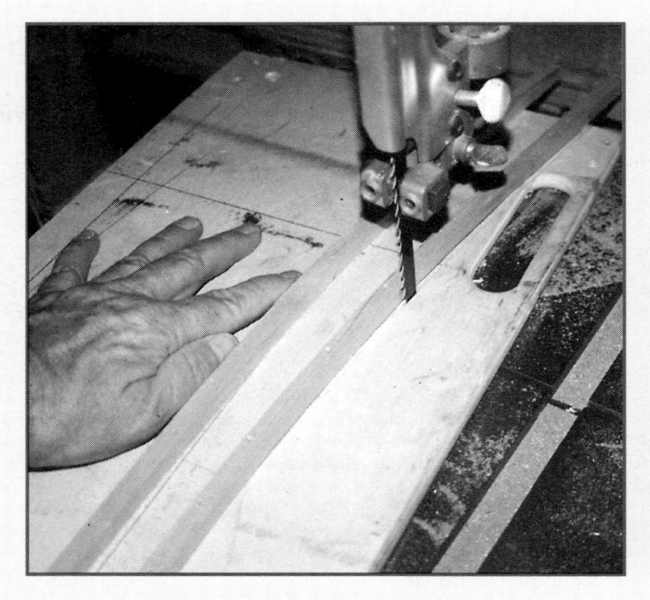

Band-saw as close to the batten as you can. The router will take care of the rest. I rarely cut with a router, preferring instead to use the router to trim off only a small amount of wood.

The bearing on the shank (pg. 87) rides on the batten to guide the cutter along the curve, making short work of shaping the plywood template. The cut will be as nice a curve as the batten without any of the normal sanding, filing, and frustration of making a curved pattern.

This is the holder for the table-leg blank, incorporating as its base the plywood template. A bearing on the shank of a flush-cutting shaper cutter works the same way the router bit does to shape the leg's curvature.

The first cut on the curved leg simply follows the line. Get as close to the line as possible, but don't saw too deeply. A fresh band-saw blade helps keep the cut under control. Remember that each new blade will have a different drift angle, as discussed in Part 2 of this book.

After the first cut is finished, tape the cut-off back onto the piece, rotate it 90 degrees, and cut the other face of the leg.

When the pieces are removed, the leg, curved on two sides, emerges from the blank.

Battens can be made of nearly anything flexible. Wood works well. Materials like this plastic are amazingly flexible.

Flat-plane curves

If you were to cut a curve out of a flat board, regardless of the shape of the cut, this would curve in one plane only. For our purposes, a curve is any line that is neither straight nor a circle.

These curves are easily cut using the batten and bearing technique discussed in Part 2. You bend a batten along the curve you wish to cut, fasten it in place with brads, then run the bearing of a flush-cutting router bit along the batten to cut the piece.

DRAWING CURVES USING OFFSETS

A curved line following no particular radius is usually drawn by connecting a series of points with a batten. In boatbuilding, this method of drawing is called "lofting." A boat builder draws, or lofts, a boat full size prior to construction. He or she does this in a loft; a large, flat area in the boat shop that is suspended above the work floor. The boat builder's guidelines come in the form of plans and drawings supplied by the designer of the boat. Part of the plans is a table of offsets.

Here is how the process works. First you draw a straight base line. Then along this base line you draw perpendicular lines each the same distance apart. In boatbuilding, these are called station lines. Then you label all the lines so they are easy to keep track of. Along each of the perpendicular lines, you measure a given distance above or below the base line. You make a mark at each measurement. Once you've made all the marks, connect the dots with a flexible batten. You draw a curved line with the batten as a guide.

Using this technique, curves can be described and communicated quite easily. Each line of a plan for a boat will have its own offsets. This is how the designer communicates the shape of the boat to the builder.

Using a batten

Let's draw a simple curved line using the chart on the following page.

Start by drawing your base line along a board or piece of shelf paper about 12 inches wide by 8 feet long. Run a pencil line the full length an inch or so in from the edge of the board. Now draw 13 perpendicular lines across the board on 6-inch centers. As in boatbuilding, these are called station lines. Label them stations 0 through 12. The process is like a cross between using graph paper and drawing a child's dot-to-dot picture

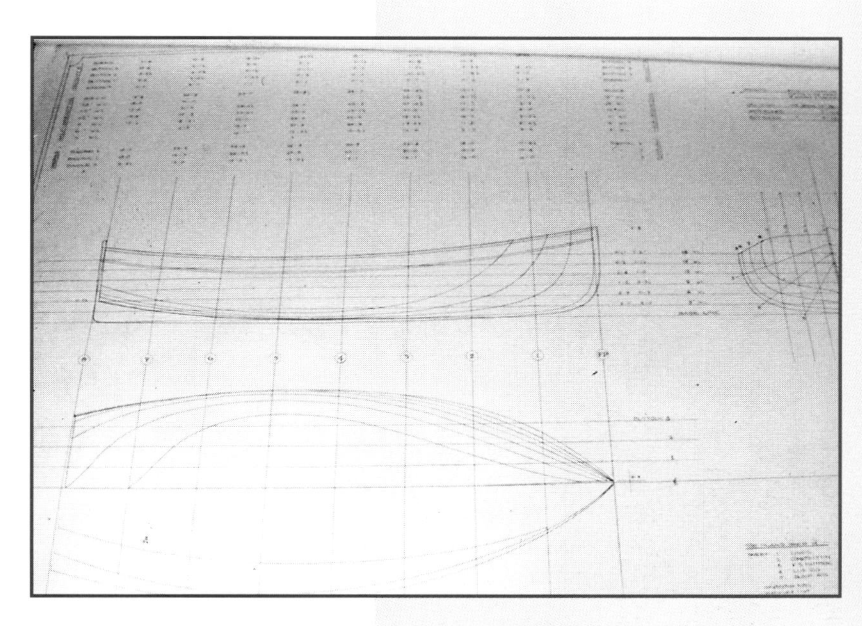

This is a table of offsets that comes with boatbuilding plans. It describes the locations of dots on a grid. Connecting these dots with a batten creates the characteristic curved lines of boats.

Now take a look at the chart. Each station has a number representing the distance from the base line to the mark on the station line—the distance the mark is "offset" from the base line. Using the chart, make a mark along each station line.

Now cut a 9-foot batten out of clear, straight wood—molding material is a good choice. Make sure the batten is stiff enough to hold its shape but flexible enough to make the sharpest part of the bend (in this case about 1/2 inch square). If the batten is not perfectly straight to begin with, don't worry. You're going to be bending it anyway.

When you use a batten to form a curve, bending it so that it exactly hits all the points may cause lumps or bumps in the line. In other words, the curve will not be fair. You must then adjust the batten so that the curve is smooth even if it means missing the marks slightly. The important thing is that the line is fair.

Bend the batten along the marks. Use small nails placed on either side of the batten to hold it in place. (Boatbuilders use lead weights with points, called ducks, to hold their battens.) Go to the far end of the board. Sight along the batten. Can you see if it is fair? If you followed the marks exactly, you will see a slight hollow around station 8 and a hump at station 9. Remember, the important thing is to make sure the line is fair even if it misses the marks slightly.

Station 8 was deliberately off by 1/2 inch to drive home the lesson that sighting along the batten from the end tells you immediately when something is wrong with the curve. Try fairing the batten by eye. You'll find that in order to do so you'll have to move the batten 1/2 inch farther from the base line at station 8. This will make the curve look better.

You can cut your parts using this same batten technique (see Part 2 of this book). The batten, if nailed securely to the board, can act as a bearing guide for a flush-cutting router bit.

Station	Offset	
0	1-1/4	
1	1-5/8	
2	2-1/2	
3	4-1/8	
4	6-1/4	
5	7-7/8	
6	8-1/2	
7	7-7/8	
8	5-3/4	----- (6-1/4)
9	4-1/8	
10	2-1/2	
11	1-5/8	
12	1-1/4	

The chart of offsets lets you plot the curve. You can tell if the chart has mistakes by sighting down the batten. Here the chart was off at station 8.

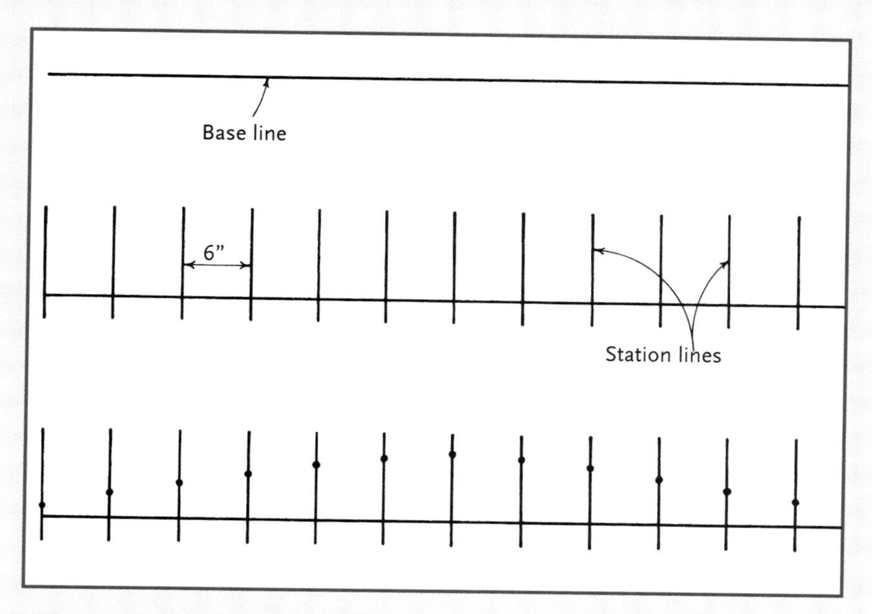

After marking the offsets onto the station lines, you can spring a batten to form a fair curve.

By sighting down from the end of the batten, you can see when the curve is fair.

PART TWO

Milling Parts from Solid Lumber

Milling Parts from Solid Lumber

A CAUTIONARY TALE

I was alone at the shop. It was late. I was running a batch of small curved handrail fittings when suddenly the piece I was working on slipped out of its jig. WHAM! The sound of the shaper cutter slamming into the end grain of the piece of oak could probably be heard a block away. The face of the cutter had impacted the dislodged piece of oak and its rotation was completely blocked. I could hear the motor running even though the cutter was stopped. Was I ever glad I had loosened the belt just for this purpose. I held on with both hands to the jig handle. Ever so carefully, I reached out my left foot and hit the paddle on the off-switch. Was I ever glad I had built the paddle switch. The machine shut down. I breathed again.

The bad news is that I made a jig that failed to hold the piece securely. The good news is that even though the jig failed, the rest of the safety features I had built into the process protected me from harm.

This illustrates why jigs are so important to milling. It also illustrates that careful attention must be paid to how the jig is constructed. It was after this particular incident that I instituted the "screwdriver" rule in my shop—that is, if a piece could be pried out of its jig with a screwdriver, it was not being held in the jig securely enough. If I had used this test before making that particular cut, I would have discovered that the piece was an accident waiting to happen.

Remember, any time a machine is spinning a cutter, there is potential for an accident. Jigs and fixtures will add safety to an operation as long as they work well and are well designed. Machine adjustments will help as well. Safety tooling will minimize the dangers involved. But the risks will remain. Please take the few minutes required to read and think about the information in the Machine Safety Appendix. ■

AS OFTEN AS NOT, milling is the first and best choice for making curves. Jigs and fixtures make the work relatively safe, fast, and accurate. In my shop, I consider other alternatives to making a curve only after I have concluded that it isn't feasible to mill the part.

This section deals with advanced machinery operations, including radius cuts on the shaper. Machines like shapers are designed to be run by experienced woodworkers, people with years of experience on similar machines like big router tables. In this part of the book I introduce the notion of using a small shaper rather than a large router-table setup because so many woodworkers outgrow their router tables. Keep in mind, though, that you must use your good judgment as to which operations you feel comfortable with and which operations to avoid for now. Remember, no piece of furniture, no matter how beautiful, is worth an injury.

ADVANTAGES OF MILLING

Milling is the process of cutting parts out of solid stock. No bending of any kind is involved. You simply find or glue up a block large enough for the part, cut it out, and mill it to shape.

Milling is extremely precise and predictable, and once the work is set up, operations move quickly and easily. If you have several parts to make, each can be made identical to its predecessors. You can repeat the same shape over and over by flush-cutting with a pattern.

As long as the arc of the curve is less than 90 degrees, there is usually no better way to make a curve. The parts are extremely strong and stable, at least up to a point where the arc approaches 90 degrees.

Milling is cost-effective if labor and materials are taken together. Unprocessed lumber may be used, even of a lower grade, since you can saw around imperfections in the stock.

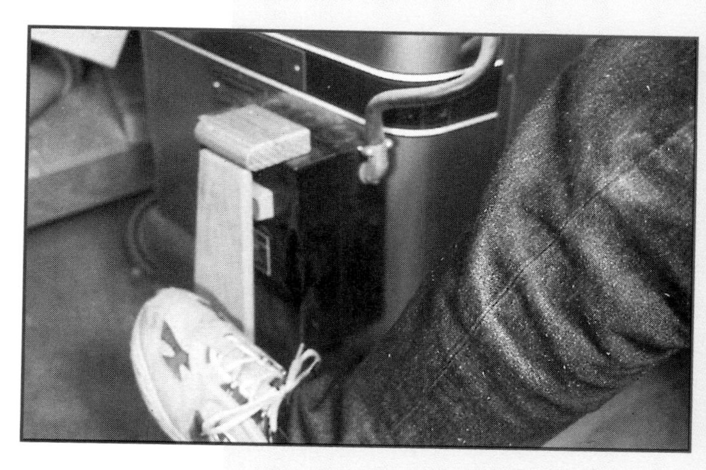

One of the first modifications I made to my shaper was to install an off-switch I could reach with my foot. This allows me to hold my position and shut off the machine without having to let go of the cut and bend down.

DISADVANTAGES OF MILLING

The main drawback of milling, after the personal risks involved, is that the strength of a milled piece falls off rapidly beyond 90 degrees of arc. This is because wood is strong in only one direction—along the grain. It lacks cross-grain strength. Those with a sadistic streak will appreciate the illustration of the strength of wood with this hypothetical martial arts demonstration. The hapless black belt approaches the pile of 1-foot square pine boards intent on breaking them with one downward head blow. Trouble is, some villainous jokester has taken the middle board and rotated it 90 degrees. With any luck, he or she has also had the foresight to alert the paramedics to stand by.

But back to the disadvantages of milling. Knowing that wood is strong only along the grain, you can see that if you were to mill a 180-degree arc out of a solid block, the piece would be extremely weak. This is because at the top of the arc the grain would be short, going across the arc from side to side. You can see this in the top photo on the next page. Any time an arc milled from solid stock exceeds 90 degrees, this problem occurs. Beyond about one-quarter of the arc of a circle, grain runout causes the curve to become weak. Joints of some kind must be used to continue the arc beyond 90 degrees.

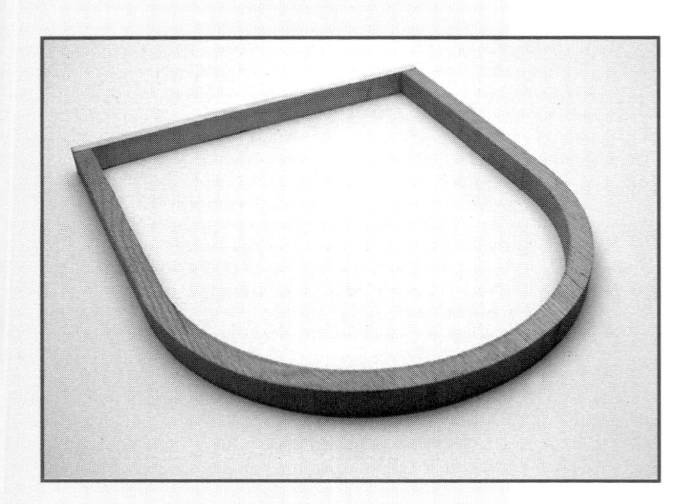

It's easy enough to band-saw an arc out of solid wood, but the resulting short grain at the top of the curve makes the piece very weak.

However, joints create both visual and technical problems. Cross-grain lines can interrupt the smooth flow of a piece. For this reason, it's best to avoid butting end grain if possible. When two boards are butted end grain to end grain, they must be carefully and mechanically joined to ensure durability.

Finally, with milling, as with all curve-making methods, a great deal of precious lumber goes up the dust chute. Careful planning minimizes waste, of course, but milling is probably the most wasteful of the three methods in terms of material usage.

SOLID-STOCK MILLING OPTIONS

There are two ways to mill solid stock. One way is to move a portable machine, like a router, in a circular or curved path across wood that's held stationary. Another option is to cut the curve in the stock by moving it across a stationary machine such as a band saw, table-mounted router, or shaper. Using either of these methods, just about any curve is possible to mill given the limitations of grain runout. Just remember that either the portable machine or the portable workpiece must be guided at all times by a sturdy jig or setup of some kind. Simultaneously, the stationary component—again, piece or machine—is securely held in one spot, usually with bolts or at least large clamps.

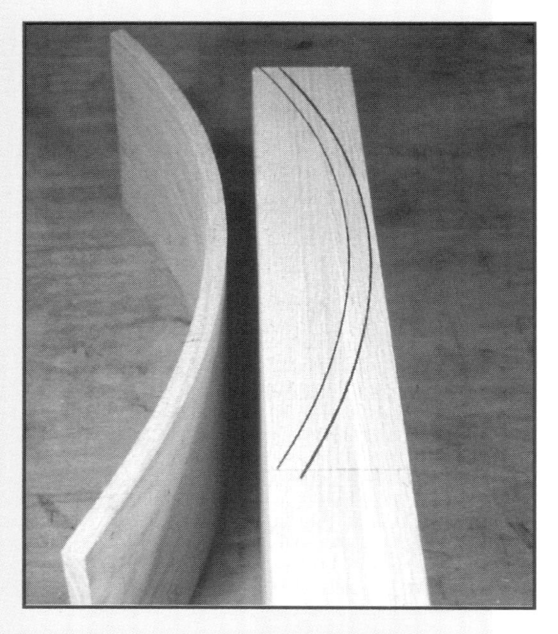

This is the size of the block it would take to saw out the crest rail for a chair, compared to the amount of material that would be used if the rail were steam-bent, as discussed in Part 4.

The choice between moving the machine across the wood or the material through the machine is influenced by the size of the part being milled. A large part, such as a circular table, is more easily held stationary while a portable machine cuts. By contrast, the table's perimeter molding, for example, would be more easily moved across a stationary cutter. Obviously, your choices will be shaped by the machinery you own, but in general, the smaller object—whether it's the machine or the workpiece—should be what gets moved.

ESSENTIAL TOOL—THE BAND SAW

Even if you're a dedicated hand-tool fan, cutting curves out of solid stock without a band saw is a miserable task, especially if you have a lot of parts to mill. A good band saw (or even a not-so-good band saw plus a little money for improvements), makes short work of cutting curves. Typically, the band saw is used to rough-cut parts before smoothing with another machine. The closer to the line you make the cut, the less cleanup you'll face later.

Cutting curves on the band saw

Cuts on a band saw may be made either with a guide or freehand. Either way, there's an inherent band-saw quirk that's sure to affect your setup. It's called "drift," and refers simply to the fact that each saw blade has its own natural cutting angle. For an accurate cut, the fence, if you use one, must be placed at the same angle. To test drift, take a board as thick as the one you are about to cut and draw a straight line down the center. Cut freehand along the line, doing whatever it takes to keep the cut along the line. When you begin the cut and you're trying to stay on the line, the blade will inevitably wander off to one side. Turn the piece so that the blade cuts back to the line. Chances are you will overcorrect. Like steering a boat, turning the wheel this way and that to hold a straight course, eventually you'll find the angle at which the blade will stay right on the line. Cut about halfway along the length of the board. Now stop the saw without moving the board. Chances are that the board is at an angle to the saw table, not parallel. Even though the cut is exactly along the line, if you don't move the board when you stop the saw, the board will be at an angle to the table. You've discovered the drift angle. The angle of the test piece is the natural cutting angle of this particular saw and this particular blade. Any curved setup, especially a guided cut, must take this angle into account. Draw a line along the angle of the test piece. This is the reference line—the drift angle—you'll need for setup purposes.

RULES OF THUMB FOR MILLING SOLID LUMBER

Make a test cut on scrap of the same type of wood as the real thing. This will help determine how well the setup works.

Make an extra part or two, especially if the setup involves more than one step. If you make a mistake, it will take a lot of backtracking and repeat setups to produce another part.

Complete all milling operations at one time when possible.

Make light, multiple cuts rather than a few heavy passes.

Make stout jigs that don't allow the workpiece to be pried out with a screwdriver. ■

Make a freehand cut in a block of wood and the drift angle will become evident. Draw a line down the piece that's about as thick as the one you'll be cutting. Cut down the line about halfway, then hold tight and stop the saw. The drift angle is the angle that the board must follow for the saw to hold to a line.

RULES OF THUMB FOR FREEHAND CUTTING

Never let your fingers cross the line of cut.

Practice on scrap.

Use a fresh blade. I routinely change blades every day or at least any time I have to make an important cut.

Make escape cuts for tight radius work.

Cut shy of the line, but as closely as possible, leaving only a slight amount to remove with the cutter.

Move the piece forward at the same time you pivot it. ■

BAND-SAW RULES OF THUMB

Accurate cutting requires a well-tuned saw.

The bigger and more powerful the saw is, the better.

The blade must be sharp, with the correct width and teeth per inch for the project.

The narrower the blade, the tighter radius it will cut. The wider the blade, the easier it will follow a line.

A day spent tuning up the saw will pay dividends for years. Even with new saws, tune-up is often required. Align the wheels, make sure tires are round and not lumpy, and adjust blade guides so the blade tracks properly. Wheels must be balanced and concentric. There have been many books and magazine articles on the topic—refer to them as needed.

Blade guides are essential. Stock guides are often deficient, allowing the blade to wander. After-market bearing guides will improve the performance of nearly any band saw.

Even high-quality, tuned saws will "drift," or not cut parallel to the fence, as explained on the previous page. This fact of life takes getting used to, but once understood, it makes setting up a saw very straightforward. ■

LET THERE BE LIGHT

The process of band-sawing will be safer and more accurate if you provide a dedicated light source. The older I get, the more conscious I grow of insufficient light. A desk light mounted with a clamp so it shines on the work area will improve your vision and thus the quality of your work. ■

Freehand cutting

Most of the time I cut curves on the band saw freehand. If you're going to do this, practice on scrap first—choose scrap that's the same material and thickness as your project. By the time you get to working with the real thing, you'll be warmed up enough to cut close to the line without veering into the part you want to keep. Practice until you are confident in your technique and realistic about how close to the line you can cut.

Here's a cautionary word on lines: If you change your layout, erase the original line. When sawing, you want to be certain that you are cutting along the correct line, so a single dark line is the best guide. Mark the waste side clearly.

Band-saw blades

A narrower blade with fewer teeth and greater set will cut a radius easier than a wider blade with more teeth and less set. A 1/4- or 3/8-inch wide skip-tooth blade will cut all but the tightest curves.

The thicker and harder the wood, the more difficult the cut will be. The fewer teeth on the blade, the thicker the wood you can mill. The space between the teeth allows a place for waste to collect. The thicker the wood, the more important this becomes.

Blade tension is extremely important when cutting thick boards. The greater the tension, the more perpendicular and straighter the cut. A bowed cut can result from a blade that has too little tension. This is another example of the advantage of larger, higher-quality tools—with a larger saw, greater tension is possible.

Guided radius cuts made with a circle-cutting jig pose the same problem as guided straight cuts—because of drift, the band-saw blade may not cut a true line to the table. Make sure you know the natural cut angle of your particular saw. This angle is what you use to locate the center point for a radius cut.

A simple circle-cutting jig

With one of these jigs, you'll be able to cut multiple duplicate circles quickly and easily since you won't have to do any layout. The idea is basic: An adjustable jig holds a single pin a set distance away from the band-saw blade. When the workpiece is in position on the pin, the piece rotates and the saw cuts the circle.

This jig is a variation of the literally hundreds of jigs designed to cut circles on a band saw. As mentioned, begin by deciding the range of circle sizes you want to cut. Then find the drift angle (pg. 59) of the saw blade. Next, cut a piece of plywood large enough for the jig. Mill slots for the hold-down bolts using a straight bit in a router; a smaller bit cuts all the way through the piece while a larger bit cuts only partway down, allowing the bolt head to lie just below the surface of the jig. Grind the head off a slotted wood screw, leaving the slot but removing the head to the diameter of the shank. This will be the pin. The remaining slot will be enough to drive it into the jig as a pivot pin. Cut a slot for the blade as long as the slots you cut for the hold-down bolts. Finally, carefully mark the locations for the two hold-down bolts by positioning the jig on the band saw. Drill and tap holes for the bolts as discussed at right.

This simple movable pivot will cut an accurate radius every time. Drill a pilot hole in from the edge of the workpiece that measures the length of the radius. The more adjustable you make this jig, the more useful it will be. Try to figure the range of circles you'd like to cut. The jig in the photo above will cut a circle as small as 4 inches in diameter and as large as 20 inches in diameter. The machine's capacity finally limits the size of the radius.

DRILLING AND TAPPING HOLES

I've made dozens of jigs that rely on drilled and tapped holes for attachment. The marvelous thing about cast-iron table tops on stationary tools is that you can pretty freely drill and tap holes for holding jigs and fixtures. It does no harm to have a few tapped holes in a cast-iron top.

No amount of clamps is going to give you the convenience and security of a bolted jig connection. Having said this, be careful to locate the holes so they don't pierce the vertical webs in the underside of the casting. These webs reinforce the table top and help keep it flat. Feel around the casting—you'll easily locate the webs (see pg. 62).

It's quite possible to do all of your drilling and tapping in just a few sizes, for example, 1/4"-20, 3/8"-16, and 1/2"-13. The first number is the diameter of the screw; the second is the number of threads per inch. These are common screw sizes, available in lots of different lengths. Besides the taps, you'll also need a tap wrench, cutting oil, a center punch, and drill bits that match the tap sizes. Most home centers sell a tap and drill as a set in a blister pack. This ensures the proper size drill for a given tap. Store the two together and the set will be ready to go.

Locate the holes and punch with a center punch so you'll be able to start the drill in just the right spot. Use light pressure to drill the cast iron. Drill the holes as straight as possible. Use regular twist drills, not brad points. Place the tap in the wrench, drip on some cutting oil, and start the tap in the hole.

To get the tap started, hold it straight as you turn it. Once the threads take hold, it's easy to keep the tap going. Don't turn the wrench fast or hard. Instead, gradually ease it around, backing off frequently. Go forward half a turn, go back half a turn, and so on. This allows the waste metal to break off and emerge from the hole. Back out the tap entirely when the tap gets hard to turn, remove the excess metal shavings, drip more oil on the tap, and return to the hole. If you take your time and don't force the tap, the process will go smoothly. ■

Drilling and tapping holes

By learning to drill and tap holes, mounting fixtures to your machines becomes a snap.

Avoid drilling through the webs on the underside of cast-iron castings. They reinforce the casting.

Taps and drill sizes are matched. Make sure to use the proper drill size, wrench, and some cutting oil to lubricate the tap.

After marking the location of the hole, punch it with a center punch. This ensures that the drill will start easily where you position it, without skidding.

An ordinary, but very sharp, twist drill will easily go through the relatively soft cast iron.

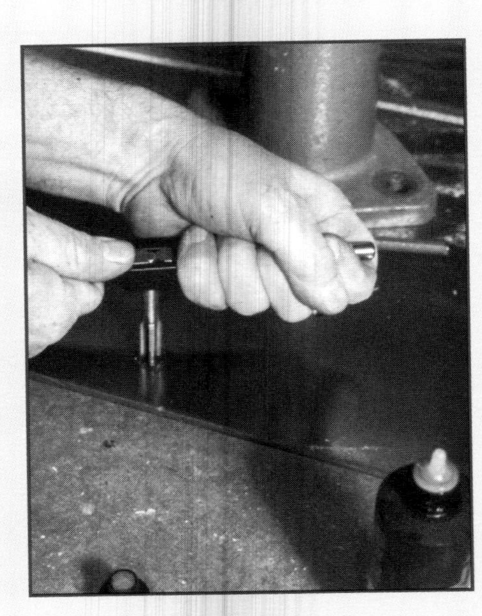

Start the tap as straight as possible, using plenty of oil. When it starts to turn hard, back it out and clean out the chips. Yes, it's a bit messy, but the security of bolting fixtures in place instead of clamping them is well worth the effort.

A partial-radius cutting jig

The pivot point of this jig must be solidly bolted to the saw. For a relatively small radius, you can use the simple circle-cutting jig shown on pg. 61. If the segment is larger, you'll have to rig a setup that looks something like a section of a wagon wheel with two spokes. The spokes hold the piece being cut the right distance from the center. The center itself is located the proper distance and angle from the blade to prevent wandering. (See the discussion of band-saw drift angle, pg. 59.)

If you are cutting out a section that will have both a convex and concave cut, you will need to make two different jigs. On the outside cut, the blade simply enters the wood as it would with a normal cut. The wagon wheel section applies here. On the inside cut, the concave part, the challenge is to make the cut without sawing the jig in half. This seems like a joke, but it can be tricky. For one thing, there has to be a recess in the jig for the blade. The piece has to be free of the blade while the machine starts. The piece has to be fastened into the jig around the blade. I use a couple of screws to do this. Then the cut can begin.

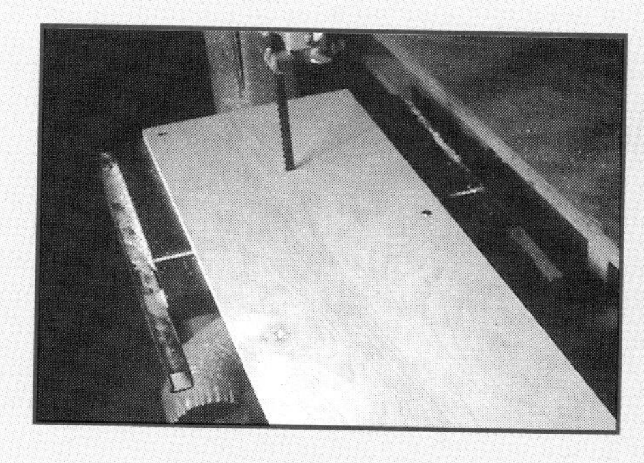

By using drilled and tapped holes to attach an auxiliary table to your band saw, you can rig an accurate circle cutter, placing the pivot point as far from the blade as needed.

Here the jig is set up for a 30-inch circle, but nearly any radius can be cut. by extending the spokes and the table.

The inside cut is tough since it is easy to cut right through the spoke of the jig. I go slowly toward the end. If I have a lot of pieces to cut, I use a stop block.

This fun jig holds the piece at an angle while it pivots around a radius, resulting in wood cut to a helix shape.

It's more straightforward to do this on a large radius by fastening an auxiliary table to the band saw through tapped holes. This ensures that the center point will be solidly attached to the machine.

Helix-cutting jig for the band saw

You may never have use for a jig like this, but it illustrates the amazing shapes you can cut with the band saw and a jig. It can be a lot of fun to dazzle people with cuts like this.

This curve is cut on a radius and also a slope. The piece being cut is held on a slope while the entire jig rotates. To build the jig, determine the slope and inside and outside radii. Make the jig as pictured. The piece being cut must be supported independently on each side and underneath. Once the cut is made, the piece must remain supported. This is one of those cuts that looks impossibly complex but in actuality is quite simple.

Making the cuts as if it were a normal radius, the tricky part is to make the jig so that it can be sawn into basically three parts but still stay intact enough to support the part. By sawing on both the inside and outside radii, you essentially saw the jig into pieces. For this reason, the construction is tricky, but the result is amazing. If you don't yet own a brad nailer, this jig will probably drive you to go out and buy one.

This piece has a helix shape cut on the band saw using the jig shown.

ESSENTIAL TOOL—THE ROUTER

There is no more versatile machine for making curves than the router. This is one of the very few woodworking machines actually designed to cut curved shapes. With the right tooling and setup, perfect circles and curves are a snap. If you are limited to only one shaping machine, the router is a good choice.

Most routers have a hole in the base to accommodate rods that can be fastened to a pivot point. As the router rotates around the center point, it will produce an almost perfect circle. Circle-cutting jigs mount in the router base holes.

Full-circle cutting jig

In figuring out how to build a commission for a large round table a few years ago, I had to make a choice. Would I put up with a small flaw in the table top (the bird's mouth indentation shown at lower left), or would I do whatever it took to have the top turn out perfectly? You guessed it—I built the full-circle cutting jig described here, which allows a straight router bit to enter the cut from the side, not plunge in from the top. I introduce the spinning bit to the cut gradually as the router goes forward around the circle. A T-handle with a screw thread pushes the router toward the center of the circle, sliding the router along the threaded rods, until the desired cutting depth is reached. Then I lock down the router for the duration of the cut. When the router returns to the beginning of the cut, it's gradually backed away until it clears the freshly cut edge. The result is as close to a perfect circle as any woodworker has the right to expect.

Using this jig, you can also cut a segment of a circle with a router. Fasten the work securely to the bench. Fasten the center point to the bench. Operate as with a full circle, only this time there is no need to gradually enter the cut from outside the circle in toward the center like with a full circle. When cutting a partial circle, you can enter the cut from the end of the piece.

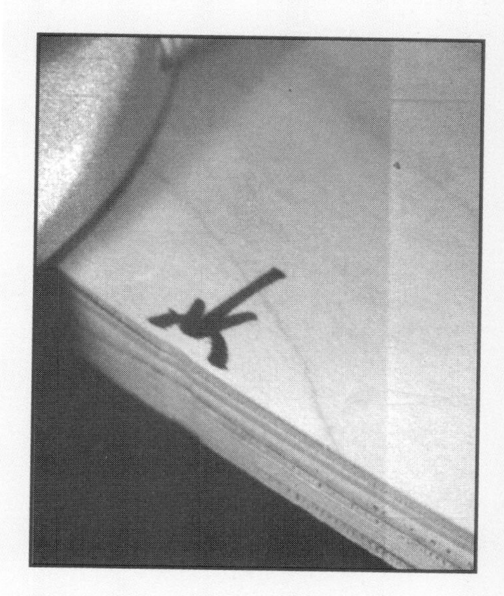

This aggravating indentation is a downside of using conventional circle-cutting jigs. I call it a bird's mouth, and it results from plunging the router into the cut. When the bit inevitably lurches toward the center, it leaves behind an area of deeper cut.

This simple mechanism moves the router gradually in toward the center point. After you start to move forward, smoothly enter the cut, complete the circle, and reverse the process to exit.

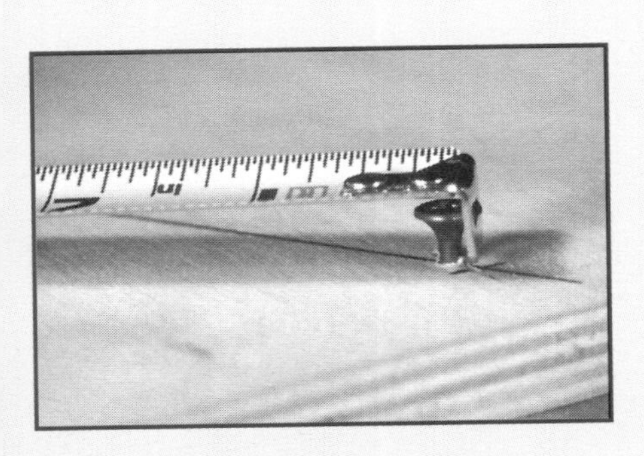

Ever wonder what the slot in the tape measure hook is for? It's for drawing circles. The slot goes right onto a screw or nail for a pivot.

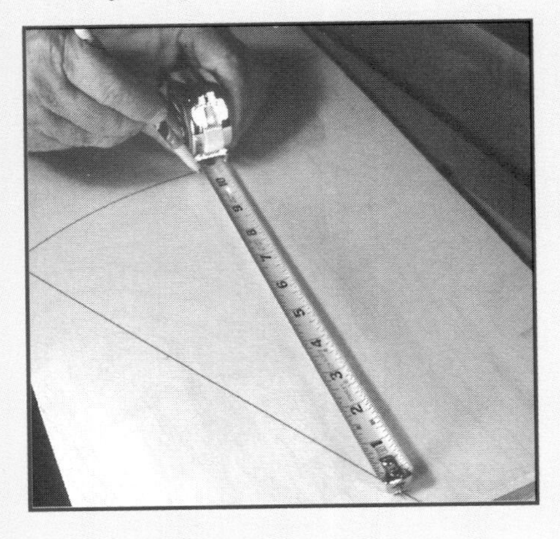

You can lock the body of the tape measure wherever you wish to draw a circle. By holding a pencil against the tape, you can then draw a perfect radius.

Constructing the jig—This jig is made of metal. It consists of several parts that are readily available from most hardware stores or home centers. First are the threaded rods, which I use because they make just about any radius easy to adjust. Using nuts on either side of the pivot and adjustment mechanism, the entire assembly is easy to build and work.

Next comes the pivot. I use a piece of angle iron most of the time, but a block of wood works just as well as you will see.

The following section assumes you have a circular table top to cut. For this example, let's say the diameter is 4 feet. If using solid wood, glue up a slab big enough for the top. If using plywood, cut a sheet in half to get a 4-foot square piece.

Draw a circle on the piece with a pencil and trammels. If you work from the bottom, you can drill a pivot hole in the center. By placing a nail with a small head in this hole, you can draw the circle using your tape measure. The hook should have a slot in it for this purpose. Lock the tape at the correct radius and hook the pencil on the body of the tape to draw the circle. Now cut as close as you can to the line with a band saw or a jig saw.

The router perches halfway on the table top. It's attached to the jig with two 3-foot pieces of threaded rod. I have an elderly Makita router with 1/2-inch diameter holes, which accommodate 1/2-inch diameter rod. These are about the biggest holes I've ever seen, and the heavier rod they allow makes jig operation more solid. Using coupling nuts, you can extend the threaded rod as long as you like. A coupling nut is a long nut used to connect two pieces of threaded rod. You could literally connect sections of threaded rod as needed for the radius you are working with.

Alternatively, you could insert a section of wood with threaded rod at each end to make a larger radius. Make two more pivot blocks. Attach them to the end of the threaded rod. Now bolt these blocks to a section of wood. A 1x4 works well for this. While there is no theoretical maximum for the radius, the jig gets wobbly once you exceed a 5- or 6-foot radius. This causes the router to lurch rather than run smoothly around the circle.

Shop-built circle-cutting jig

Here are all the components you'll need to make your own circle-cutting jig—some threaded rod, nuts and washers, wing nuts, T-nuts, a couple of hardwood blocks, and your favorite router.

Mark the locations for holes for the threaded rod, pivot point, and the T-nut for the crank. Drill the holes oversize so the blocks can slide along the threaded rod.

The router slides along the threaded rods. As you turn the crank, you very gradually move the router into the cut. Once it's cutting the way you want, tighten the lock nut on the router base and continue around the circle.

The pivot works by anchoring the jig to the center point. The various ways to anchor it to the project include temporarily gluing a block to the center and turning the project over so you can drill a hole in the underside where it won't show.

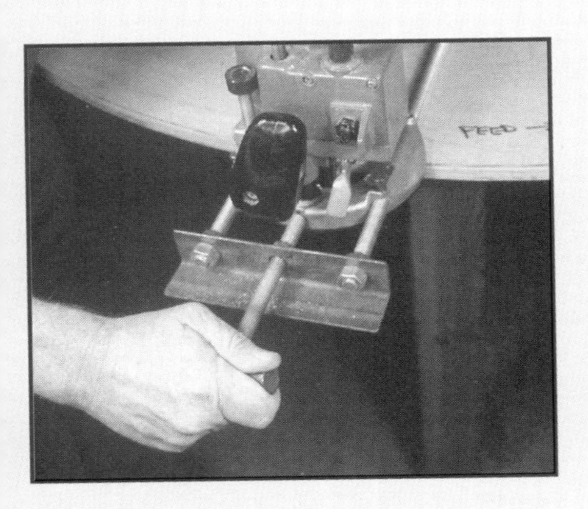

Twisting the screw moves the router ever so gradually toward the center of the circle, engaging the workpiece laterally instead of vertically, which eliminates the bird's-mouth problem.

The screw adjustment moves the router along the threaded rods so the bit gradually enters the work. It's the key to cutting perfect circles. Make the screw adjustment by mounting a T-nut in a wood block that is bolted to the threaded rods with nuts on either side. If you can manage to attach a handle to the bolt somehow, the jig will be less awkward to operate. I welded a short section of pipe to the bolt, to make it easier to turn without needing a wrench.

The pivot is nothing more than a block of wood attached to the rods with a nut on each side. A wood or lag screw down into the center of the table should have just enough clearance in the block to allow the jig to rotate. A flat washer inserted between the pivot block and the table top will make the jig operate more smoothly.

With any circle-cutting jig, there has to be a firmly anchored pin at the center point of the circle. There are two ways to do this without having a hole exactly where you don't want one. It is essential that this pivot be as rigid as possible. This single screw will be totally responsible for guiding the router.

If it's possible, turn the table top good face down and simply run a screw into the underside. (It goes without saying that it's a good thing to be careful in drilling the pilot hole so you don't drill through the top. A small piece of masking tape stuck on the drill bit is what I use when I want to limit the depth of a drilled hole.) Since you are building a circular shape, chances are you know where the center point is, but if you are working from an existing circle, simply measure across to get the diameter. Divide this by two and make a mark. Now measure in from the perimeter of the circle in several spots, making a mark each time. You may not have marks in the exact center, but they will be grouped near the center. Make a cross in the middle of the group and you have the center.

If for some reason you must work from the finished face of the top, take advantage of a technique common in faceplate turning. Glue a wood block to the center of the table top using a piece of heavy paper in between the top and the block. Now you can screw into the block at the

ROUTER RULES OF THUMB

Buy the highest quality router(s) you can afford.

The more features a router has, the better. Increased flexibility means increased problem-solving ability.

Larger-shank bits (1/2 inch) will perform better than smaller, 1/4-inch shank bits. The larger shank means less vibration and fewer chatter marks.

A separate collet for different-size shanks works much better than the spacer that some routers still use.

Collets must be tight so bits don't wander out of their collets.

Light, multiple cuts are safer and more accurate than heavy passes and leave a better surface.

Don't use large-diameter bits without a substantial router table. When in doubt, move up to a small shaper equipped with a router-bit collet. ■

center of the circle for the pivot pin. When you're finished, separate the block from the table by inserting a stiff putty knife in between. This will tear the paper, allowing the block to come off. Now the paper will have to be scraped off. Work thoroughly and carefully—leave no glue residue or it will effect the final finish.

For gizmo fans, there is a very cool router jig that solves the problem of fastening the pivot in the center of a circle. It uses a suction cup to hold the pivot to the circle while you run the router around the perimeter. See the Appendix for a source of supply.

This router jig solves the problem of fastening the pivot in the center of a circle. It uses a suction cup to hold the center pivot in place while the router is run around the perimeter of the circle. Its precision allows you to cut an inside and outside circle that fit together.

The suction comes not from a vacuum pump but from compressed air pressure. Hook up the hose from your compressor and the pivot sucks down to the surface so powerfully it's practically immovable.

ESSENTIAL TOOL—THE ROUTER TABLE

If you have a router, you should also have a way to mount it securely in a table so that you can move the wood instead of the machine. Once you've used this setup a few times, you will probably come to rely on it.

By making a router table out of the side extension of your table saw, not only do you save space in your shop, you can use the table saw's fence for the router.

Table-mounted router fixture

Any router can be mounted to a table. Any table can hold a router, even the side extension table of a table saw. (In fact, if you set it up right, your table-saw fence can serve double duty as a fence for your router table, as long as you clamp a sacrificial wood strip to the table-saw fence before routing.) Another frequently overlooked router-table surface is the workbench. If yours has some clear space, consider converting it into router-table surface.

In principle, conversion is simple. You cut a square hole in the surface large enough for the router to drop through. Then you buy or fabricate a new baseplate for the router out of plastic, metal, or plywood. You cut a rabbet around the perimeter of the hole, matching the depth of the rabbet with the thickness of the new baseplate. The router drops into the hole so that the new baseplate fits flush with the bench.

Start by laying out the size of the hole for the router. Then mark out a square about 1 inch larger all around than the size of the new baseplate. Cut the hole in the bench top with a jig saw.

Rout a rabbet around the hole. Set the bit for the depth that matches the thickness of the new baseplate material. Make a track for your router by first measuring from the outside cutting edge of the bit to the edge of the baseplate. Set four track pieces at this distance away from the edge of your layout lines. This way, the router will cut a rabbet right along your layout lines.

RULES OF THUMB WITH A TABLE-MOUNTED ROUTER

Get used to the idea of making a lot of holding jigs. Don't try to hold a piece of wood smaller than about 1 foot square without a jig.

Before you rout it, soften sharp edges of the wood with sandpaper. Sharp edges may cut you if the piece lurches in your hands. Softening the edges will help prevent this.

Setups work better if there are dust-collection provisions.

If a bit feels and sounds like it's too large for the machine, it is too large for the machine. Consider upgrading to a shaper.

The larger the bit diameter, the slower the rpm speed.

Always try out a new operation on scrap. ■

Table-mounting a router

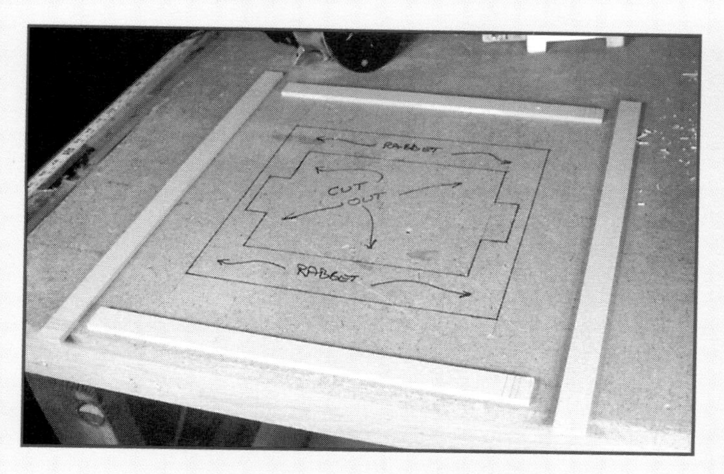

Lay out the location of the rabbet cuts and temporarily attach wood strips to the table to guide the router.

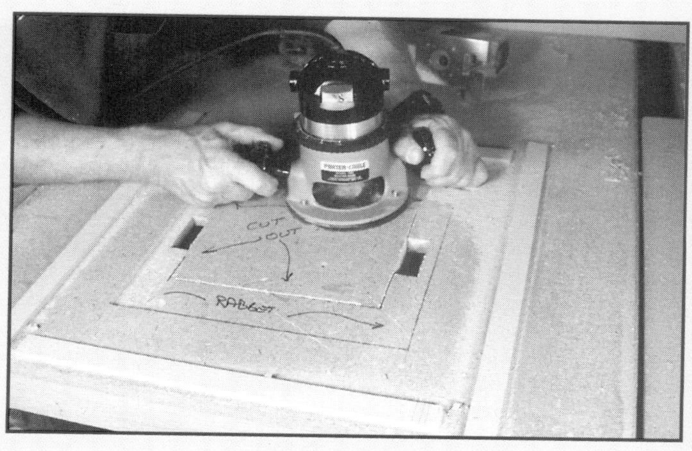

Run the router around the perimeter using a straight bit. Make the rabbet as deep as the new baseplate material is thick.

Cut out the interior of the opening. Make sure your largest router will fit through the hole.

Screw the new plywood baseplate to the router. Use the original baseplate to mark the locations of the holes. The base can be made of plastic, plywood, tempered hardboard, or metal.

The router drops right into the opening, plywood baseplate fitting into the rabbet so the surface is flush with the surrounding bench top.

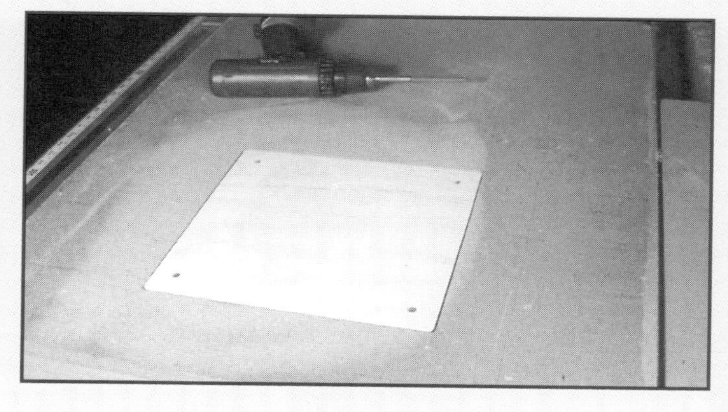

You'll want to use an extra blank baseplate to plug the hole when you're not routing.

Remove the existing baseplate from the router and cut several pieces of the new base material—set one piece aside to use as a plug to cover the hole when the router is not in place. Drill the rest to fit the mounting holes in the router base. Mount one of the pieces to the router, using longer screws if needed.

Fit the baseplate to the hole, drill holes for the mounting screws, drill a hole for the router bit, and mount the new plate to the router. Set the plate in the hole, letting it nestle into the rabbet. Drill and countersink holes in each corner. Install the screws and that's all there is to it. When you're not using your router, plug the hole with the blank plate.

With this setup, you can mount a center point anywhere on the table for a pivot point to use when cutting curves. More discussion on this method of milling circles will appear later.

ROUTER TABLES VS. SHAPERS

Router-table work and shaper work are remarkably similar. For instance, both share the principle that, to use the machine safely when moving wood past a cutter, the workpiece must be held securely. If something happens when using a portable router, at least you have your hands on the smooth router handles. If something happens when using a router table or shaper, your hands are on wood that is most likely sharp-edged from being freshly milled. If the piece lurches suddenly, you can sustain considerable injury, which is why I suggest lightly sanding the edges of the pieces. The essential point here is that the shaper is so much larger in scale—it has more powerful, larger tooling, and a greater potential for problems. For that reason, I suggest you do not consider working on a shaper before you accumulate years of experience on router tables. No doubt you will have a few—hopefully minor—mishaps with your routers, which will help you anticipate the similar, but much more severe, problems that can arise with a shaper.

The essential differences between a router-table setup and a shaper are that the shaper is more solid, more powerful, and is capable of cutting much larger profiles than the router. You can often tell it's time to move from a router to a shaper by your bits. The day will inevitably come when you put a large bit in your router, start up the machine—and your stomach will churn since your tool now sounds like a jet engine hooked up to wild horses. You'll quickly shut off the machine and contemplate how badly you really want to make that particular cut. What the router is probably telling you in this situation is that you've outgrown your rig. You might just be ready for a small shaper.

By "small shaper," I don't mean the kind that sits on a table top. These are not really shapers and represent, in my opinion, the worst of both worlds. With their very small (at least by shaper standards) 1/2-inch spindles, they have the same inherent problems as the router table—too small a machine for too large a bit.

The small shaper I am referring to is a cabinet model weighing about 450 pounds. I have a 3hp Powermatic Model 26, but other similar makes are available. When you go out to buy a shaper, make sure tooling is available to run router bits as well as conventional shaper cutters.

If you were to take the same bit that was too big for your router and chuck it up in a small shaper equipped with a router-bit collet, you would be amazed at the difference. The shaper, with its added weight, slower rpm, and massive bearings, is designed to make pretty large cuts, so even a large router bit is quite small by shaper standards.

With a router, the collet is attached to the end of the arbor on the motor itself. Shapers have a separate spindle, usually driven by a belt in the same way as a table saw. The motor is separate, like it is on a table saw. The spindle rotates on its own bearings and moves up and down with a crank, making fine height adjustments easy and precise.

The biggest difference between a router-table setup and a shaper is size and scale. The larger spindle in a shaper will handle cutters weighing several pounds, which can cut an entire handrail profile. The shaper spindle alone could weigh more than the entire router.

The white European cutter (at top) is made for hand-feeding. It is inherently safer than its more common North American counterpart (in the foreground), since cutting depth and opening sizes are very limited.

In some ways, though, a shaper is very much like a big router table. The shaper spins a cutter or router bit that sticks up through a hole in the table—just the way a router-table setup works. Many shaper setups will be familiar to anyone who has spent any time using a router.

The important thing to remember is that if you are not really familiar with using a table-mounted router, don't consider using a shaper. You need to build a history of experience with the router to gain a feel for the much larger, more powerful, shaper, its tooling requirements, and potential hazards.

Differences in router and shaper tooling

The tooling you use must accommodate hand-feeding most of the time. This means that safety cutters are required. These cutters are made inherently safer by incorporating chip limiters and anti-kickback design.

Developed in Europe, safety cutters limit the opening, or gullet, near the carbide cutter, limit the cut by making the body of the cutter just slightly smaller than the cut diameter, and, in some designs, even eject objects if they penetrate the cutting circle too fast.

A common flush-cutting bit for a router is a typical 1/2-inch diameter straight bit with a cutting length of up to 2 inches, about the maximum length for such a bit. It invariably cuts with chatter marks no matter how carefully the work is fed. The shank itself oscillates and vibrates as it cuts, which is what causes the chatter marks. (See the pattern-making section starting on pg. 84 for a solution to the problem.)

On a shaper, chatter marks are virtually eliminated due to the greater mass of the machine and the size of the spindle. There's a flush-cutting setup that fits on a 1-inch spindle, cuts a 3-inch diameter, and will cut up to 3 inches in height. This larger, more solid tooling makes all the difference when you look at the resulting cut—it's as smooth as glass.

ESSENTIAL TOOL—THE SHAPER

The shaper is my favorite machine in the shop and, if set up and used properly, in my opinion is one of the safest.

Mention the word shaper in a group of woodworkers and it will usually start a lively conversation. Nearly every woodworker, it seems, knows a shaper story. A personal favorite is about the cabinetmaker who made an apron out of oak to wear while running his shaper. The story illustrates how edgy even experienced woodworkers can be around this particular machine. But as I mentioned earlier, when a router is straining to its limit to make a cut, it's time for a shaper to take over.

A typical small shaper weighs about 450 pounds, has two speeds (usually around 7,000 rpm and 10,000 rpm), has a 3hp motor, and is reversible. It will have 1/2-inch and 3/4-inch diameter spindle sizes and should have a 1- or 1-1/4-inch diameter solid spindle as well as a router collet attachment. The solid spindle is used for larger cutters.

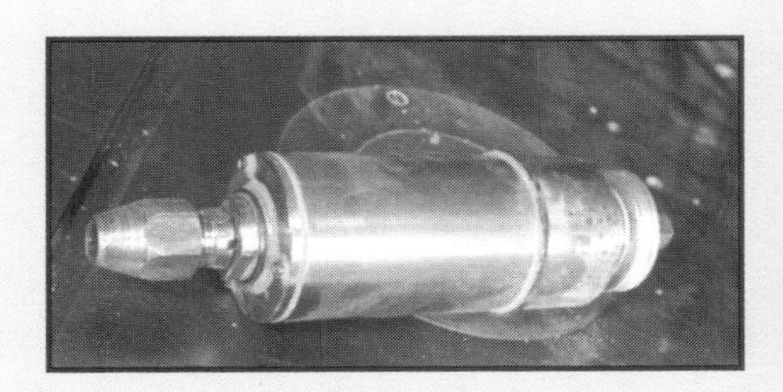

The heart of the shaper is the spindle. Weighing more than my entire router, it has massive bearings, making even a fairly heavy cut as smooth as glass.

Jigs and fixtures for the shaper

As a general rule, a jig is defined as a portable holding device used to guide a workpiece as it's cut by a stationary machine. While a fixture also guides a workpiece, it's fixed in place more permanently—hence the name. A pattern jig for cutting a curved part is a good example of a jig; more portable and more geared toward a specific shape. An example of a fixture is the horizontal-axis setup used to cut radiused handrail fittings, which you'll see on pg. 79.

A successful jig or fixture makes even the trickiest little cut a pleasure. Get used to making jigs and fixtures, since a different one will be needed for just about every surface to be cut. It does take some time to get used to the idea that you can't just set the fence and go, like when you make a cut on a table saw, but I've found that over time building jigs and fixtures becomes a rewarding exercise in its own right. Since a good jig or

A power feeder is the best safety item money can buy. It keeps the operator's hands safely away from the cut.

RULES OF THUMB WITH THE SHAPER

Don't attempt to operate a shaper without many years of experience on a table-mounted router and/or professional instruction.

A 3hp shaper is much more powerful than a 3hp table-mounted router.

The machine needs considerable setup for specialty cuts to be made safely.

The machine itself is only the starting point. It needs tooling, a feeder, holding jigs, infeed and outfeed, and much more.

Router bits may be used in the shaper as long as it has a collet for holding them.

A shaper needs a dust collector, since it produces a greater quantity of chips than a router. If the chips aren't cleared, the cutter can jam. In addition, the dust collector carries away heat, allowing the cutter to run cooler.

A shaper needs a power feeder for safety as well as for cut quality. The feeder holds the workpiece down, against the fence, and moves it across the cutter, all the while keeping the operator's hands well away from the cut. With a power feeder, even difficult molding can be produced quickly and safely.

Much of the shaper's bad reputation stems from loose knives flying out of the machine and either hitting the operator or making eerie-looking holes in the wall. Solid tooling, though more expensive in the short run, is inherently safer than loose knives simply because there are fewer things to remember to tighten and fewer things to fly apart. ■

fixture gets used over and over, the time it took to make really pays off in the long run. In the short run, never feel you have to justify spending all day to make a jig for a cut that takes two minutes. Your keywords here are safety, accuracy, and efficiency, in that order.

Use your best materials to build jigs and fixtures, like leftover Baltic birch plywood. This material will stand up to many passes on a spinning bearing without becoming worn. Besides, it justifies keeping some of the scrap that we all tend to accumulate.

Round sharp edges, carefully sand everything, and put in as much effort on the jig or fixture as you would on the project under construction. When you make a jig, try to do so with the awareness that years from now someone may walk into the shop, look up at that old jig hanging on the wall, and say something like, "Boy, isn't that thing pretty. What the heck is it for?" This brings up the notion that taking the time to make notes on the jig or fixture itself may pay off in a year or two when, try as you might, you can't remember which of your nine routers you used with the thing! I try to make it a habit to jot some notes right on the jig or fixture, and, if I'm organized enough, I fasten a sample of the cut to the jig for future setup purposes.

Some jig and fixture pointers

At the beginning of the cut, the cutter will try to draw the workpiece out of the jig. You should therefore plan to fasten the end of the workpiece with a screw, just to make sure it stays put. When running smaller pieces, make the jig a bit oversized to have plenty of room to build handles that feel comfortable.

Anytime you cut with a guide bearing, plan to make a jig with an on-ramp and an off-ramp. This allows you to stop the bearing solidly against the jig before the cutter touches the wood (see pg. 82). Then begin the cut gradually, control the cut along its length, and end it smoothly.

Jig for vertical-axis cuts

Radius work is easier than it might look, though the setup is often more time-consuming than the actual run. The setup for a large vertical-axis radius looks like a segment of a wagon wheel laid flat on the bench. The piece is run past the cutter as if it were the rim of the wheel with its spokes attached to a pivot point, just like with the band saw.

I make jigs like this from scrap plywood and lumber I retrieve from the dumpster. Not that I use inferior materials, I just try to recycle the scrap I happen to have around. Holding a piece like this is a challenge. First, I cut the ends of the pieces on the miter box. Then I cut out the blanks on the band saw. Next the two holding jigs are made to accommodate the radius and direction of travel.

The radius of the cut is measured from the side of the cutter that will come in contact with the wood to the pivot point. This is one of the reasons why it is a good idea to bolt your machine to an adjacent workbench. That way you can conveniently put a screw in the workbench at whatever radius you are cutting to secure the pivot point. The pivot point must be securely attached to the table of the shaper. If it is not secure, the cutting depth may vary, with the possible result of an irregular and hazardous cut.

The inside radius is the harder of the two cuts to set up. The cutter will contact the piece on the side of the cutter that faces away from the pivot point. Therefore, the spokes of the setup will hit the cutter if the piece is pushed too far. The simple solution to this problem is to clamp stops that will limit the travel of the spokes, to make sure that they stop well clear of the cutter.

For cutting segments of a circle, this simple jig does the job of holding the piece as it's being cut and guiding it in a circle as well. It's very important to move the piece through the cut in the correct direction. Note that the inside and outside cuts are made on opposite sides of the cutter.

Position the screw in the end grain, where the hole won't show. The screw holds the piece in the jig at the most hazardous part of the operation: the start.

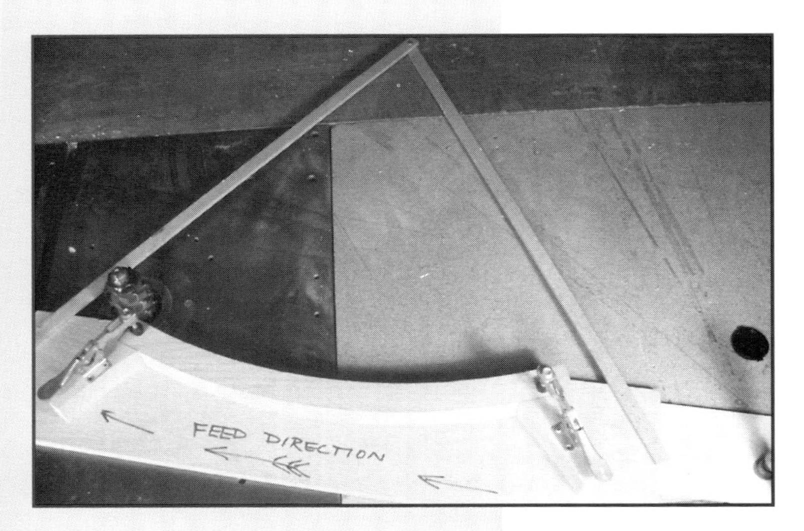

Stops and cutouts in the jig allow the cutter to spin freely before and after the cut. You can see the stop block on the left side of the jig, fastened down securely with a clamp.

Nowhere are arrows more important than when cutting curved pieces. Feeding in the wrong direction could be disastrous. If fed in the wrong direction, the piece would be caught and launched with incredible speed.

The jig will need a cutout at the beginning and end of the cut in which the cutter may spin freely. The machine is started with the cutter free of the piece in its cutout and the spoke against its stop. The cut is made until the jig hits the far stop when the cut is finished. Then the machine is shut off while the cutter spins freely in its cutout.

The single trickiest part of this operation is remembering in which direction to move the piece past the cutter. Draw large arrows on your table or on the jig as a reminder of which way the cutter spins. Once the cutter starts to spin, there is no way to know for sure—simply because you can't see it. When in doubt, shut down the machine and make sure you know.

The guard is a problem, since it has to be set up on the inner part of the radius. This is another instance when drilling and threading bolt holes in the tabletop pays off (pg. 61). The holes make it easy to bolt down a combination dust pickup/guard. In the photos, the guard was temporarily removed for photographic clarity.

The machine is started with the cutter free of the piece in its cutout and the spoke against its stop. The cut is made until the spoke hits the far stop when the cut is finished. The machine is shut off while the cutter spins freely in its other cutout.

This shop-built fixture has turned out hundreds of curved handrail fittings.

Fixture for making horizontal-axis cuts

This particular fixture is designed to make horizontal axis cuts. We use it to cut radiused handrail pieces. It works with a router bit, not a traditional shaper cutter, which was custom-made to match an original design profile used in our shop. The beauty of using a router bit as opposed to a shaper cutter is that the router bit has a much smaller diameter. This allows a profile cut on a piece with a 3/4-inch inside radius, a specialized fitting we make.

The foundation for this fixture is the vertical piece. It's made from heavy oak pieces and reinforced with plywood. The more rigid the fixture, the higher the cut quality. Vibration accounts for the ripples often seen in

USING FIXTURES SAFELY

One of the many reasons a shaper is safer and more suitable for an operation like this is that the drive belt can be loosened just enough to offer an added measure of safety. I adjust the belt to be tight enough only to spin the cutter. If the cutter gets jammed, the pulley spins on the belt, saving the fixture from destruction and allowing the operator to shut down the machine before any harm is done.

The setup for the horizontal-axis fixture has slotted holes to allow the entire fixture to slide back and forth for gradual cuts. The idea is that by moving the entire fixture horizontally, an initial light cut can be followed by deeper and deeper cuts.

A handle attached to the blank holder is used to pivot the piece past the cutter. This setup will run an inside radius as small as 3/4 inch. This cut would not be possible with a conventional large-diameter shaper cutter. As the piece rotates to make the radius cut, the cutter would distort the profile. With a relatively small-diameter router bit, however, the profile matches the other fittings almost exactly.

Note the outboard fence on the right side of the jig. This prevents the piece from being pulled out of its holder when it first enters the cut. The idea is that if the piece is allowed to move in any direction but the desired one, it is liable to be pulled into the cutter. The outboard fence, used in nearly all of my setups, prevents this, not only making the cut safer but much smoother. ■

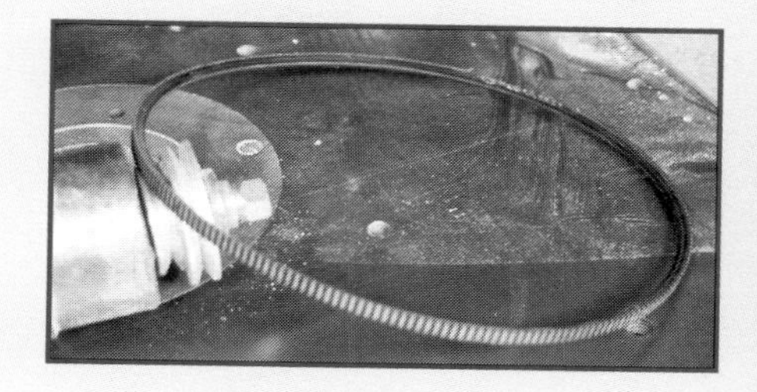

When hand-feeding, keep the drive belt looser than it should be. This allows the cutter to stop if it gets caught by something even though the motor keeps running. Tighten the drive belt for power-feeding.

millwork. The mass of the machine itself and the jigs and fixtures dampen this vibration. I made the vertical piece with slotted screw holes so it could be adjusted. I drilled and tapped holes in the top of the shaper for the studs used to bolt it down. This is certainly a case in which clamps simply lack the necessary holding power.

Glue and screw or nail the plywood braces to the oak pieces to make the basic jig. Then, using a long drill bit, drill through the center support for fastening the pivoting holding jigs.

I always try to incorporate dust collection into my jigs and fixtures. Without it, chips build up and so does heat.

The setup for a horizontal axis radius cut sits upright on the table. The pivot is suspended above the cutter at a distance equal to the radius. The setup is securely bolted to the table into drilled and tapped holes.

By bolting different tools to the basic fixture, we're able to mill both an inside and an outside radius on different handrail fixtures. Even though you may never have occasion to mill handrails, the fixture demonstrates several features of radius milling with a horizontal axis: bolting the fixture to drilled and tapped holes; adapting a basic fixture to several different operations to make different parts; making sure the fixture is safe to operate by installing handles and holding devices; and having dust collection built into the fixture.

OTHER JIGS AND FIXTURES

Toggle clamps

A great way of holding the workpiece in a jig is with toggle clamps. Toggle clamps work easily and quickly to hold and release the part. The tension can be adjusted to a setting that feels just right. But don't rely on toggle clamps alone to hold the piece in place. Use a screw or other way to keep the workpiece from being drawn out of the jig (see photo, pg. 00). Remember, if you can pry it out with a screwdriver, the piece is not being held in securely enough.

It is important to use stiff material for the base, or backbone, of the jig when using toggle clamps. The way the clamps press on the workpiece tends to bow the jig base, which will make it rock on the table.

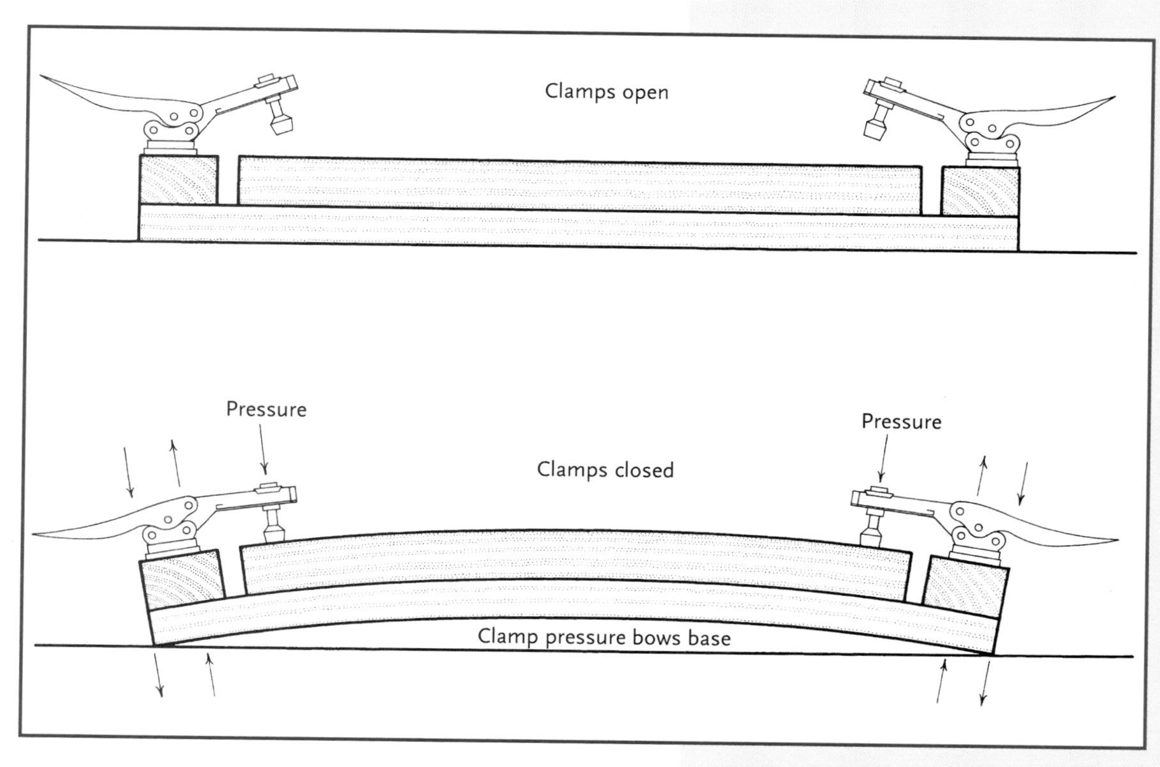

Closing the toggle clamps tends to cause the base to bow. Use stiff plywood for the base to prevent this.

Providing a place for the bearing to contact the jig before the cutter hits the workpiece keeps the beginning and end of the cut under control.

On- and off-ramps

Anyone who has used a round-over bit in a router has experienced the unsettling lurch at the beginning of a cut. This happens because the cutter contacts the piece slightly before the bearing stabilizes the cut. With larger bits, this can be extremely dangerous. Imagine this same effect with a shaper and a much larger cut. The workpiece with or without a holding jig could easily be torn out of the operator's hands. A good way to avoid this hazard is by building on- and off-ramps into the jigs you make.

The principle of on- and off-ramps is not new. Many commercial jigs have this feature. The on-ramp contacts the bearing before the workpiece contacts the cutter. The jig or pattern is then solidly grounded, with the bearing stopped before the cutter enters the piece. Likewise, the off-ramp works to keep the jig in contact with the bearing after the cutter has exited the piece.

In these two photos, you can see that this simple shop-built guard doubles as a dust-collection shroud. It could be clamped in place, but bolting it down into drilled and tapped holes makes it much more secure.

Starting pins

No discussion of jigs and fixtures would be complete without mentioning starting pins. These have fallen out of use with many woodworkers in favor of an on-ramp, but they are useful to know about just the same.

Most shapers come with a starting pin arrangement to help start the cut in a controlled way in the absence of a jig. The pin is inserted into a hole in the table a short distance from the cutter. The piece is held against the pin while the piece is rotated into contact with the cutter and the bearing. Once the bearing makes contact, the piece is then rotated away from the pin as the cutter starts work.

Most shapers have a starting pin to help start a bearing-guided cut.

Obviously, this is not an operation for the beginner if the bit is of any size at all. It is at best tricky to start a cut this way and at worst extremely hazardous. The reason it is little used is because on ramps are so much safer. I mention pins here just because using a starting pin is safer than using nothing at all.

A starting pin works well enough to use occasionally with a small round-over bit, and you can add one to your shaper if needed by drilling and tapping a hole near the cutter opening and inserting a bolt for use as a pin. Just remember, an on-ramp is ever so much smoother and safer.

This pattern uses the principle of flush-cutting. Jigs designed to be used for flush-cutting patterns need on- and off-ramps, so the guide bearing stops before the cutter enters the workpiece.

CUTTING A CURVE WITH A FLUSH-CUTTING BIT

With flush-cutting, you can duplicate the exact shape of a pattern time after time. The technique comes in handy to cut a furniture part that isn't straight, such as the rear legs/back support of a chair or a flared table leg. You'll also appreciate it in a variety of other situations—cutting a curved mantel top or a plank for a boat, for example. It's common to make a pattern for parts like this, then use the pattern to flush-cut the part.

Cutting the pattern

The trouble with flush-cutting comes with making the pattern. It's difficult to get a fair curve, to hand-work a plywood shape to a perfect contour. You quickly find that even drawing the line with a pencil is hard, let alone making the pattern out of plywood! There is seemingly no alternative to sketching the part, band-sawing or jig-sawing the pattern as close to the line as possible, and sanding or filing the pattern until it's as close to perfect as you can get it. But sanding and filing are extremely frustrating tasks, since the pattern always seems to have lumps where you cut too deeply and bumps where you should sand some more.

This section introduces a system for flush-cutting using a batten as a guide. You can use this system to avoid all the sanding and filing difficulties that come with the traditional way of making patterns.

Drawing the lines

In boatbuilding, you usually get a table of offsets to tell you how to lay out a curve (pg. 51). You develop curved lines by connecting a series of dots. You draw the curved line using a batten, or thin, bendable piece of wood. You sight down the batten to see if there are any lumps or hollow spots, moving it back and forth until the curve is smooth. While we'll use boatbuilding principles to make our pattern, we'll go one step further and use the batten to guide a router through the cut.

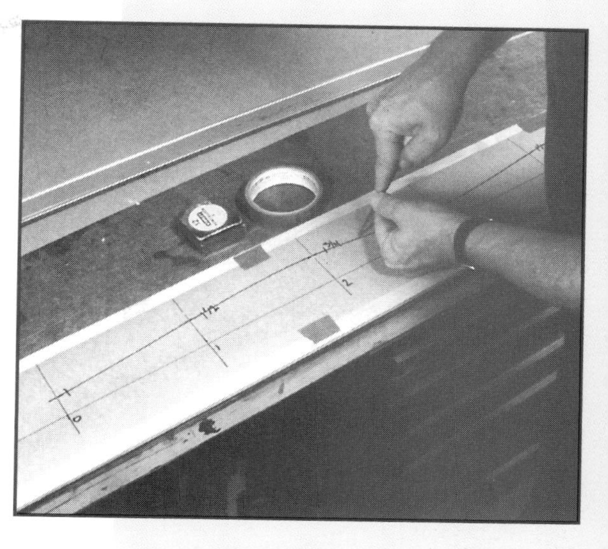

One way to transfer a curved shape from paper to wood is by pricking through the paper with an awl.

Layout with battens

Using a batten makes drawing lines easy—as long as you have a pneumatic brad nailer, an essential tool that makes this process a snap. From your design (Part 1), you'll have either a grid pattern or full-sized drawings of the part you need. It's a simple matter to prick the lines every inch or so to transfer the contour to the plywood drawing. These dots or pricks on the plywood will be the guides to the curve. Don't bother drawing a pencil line right now. You won't need one.

Mill enough batten material from clear straight-grained wood (whatever species you have around the shop is fine) so that you have several pieces. Milling all the batten material square in the surface planer ensures that all the pieces will be exactly the same size. Keep making the batten thinner until it easily bends around the contour.

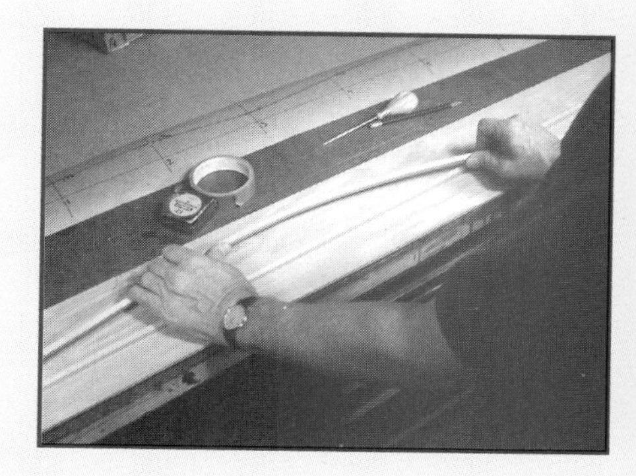

Mill the batten to whatever size it takes to keep the batten as big as possible, so the curves stay fair, but as small as necessary to bend easily around the smallest radius.

Set the edge of the batten along the beginning dots on the plywood and nail it with a brad about every inch or so, bending the batten along the dots as you go. Adjust the batten until you are satisfied that the curve is fair. Spend some time here until it looks perfect. Position the batten exactly along the line. You'll find that with a nailer, this job goes quickly. If you should split the batten or should it crack while you're bending it, it might be too thick. Pull the nails, mill the batten material thinner, and start again. You want the batten as stiff as possible without having it break as it bends around the contour.

Using the brad nailer, fasten another batten the same size, about 2 inches inboard, away from the cut and roughly parallel with it. Check to see that the inner batten is close enough to the guide batten to provide a track for the router to ride on. With a jig saw or band saw, cut along the batten as close as possible without cutting into the batten. Now, instead of

Nowhere is a pneumatic brad nailer more valuable than when fastening a small batten.

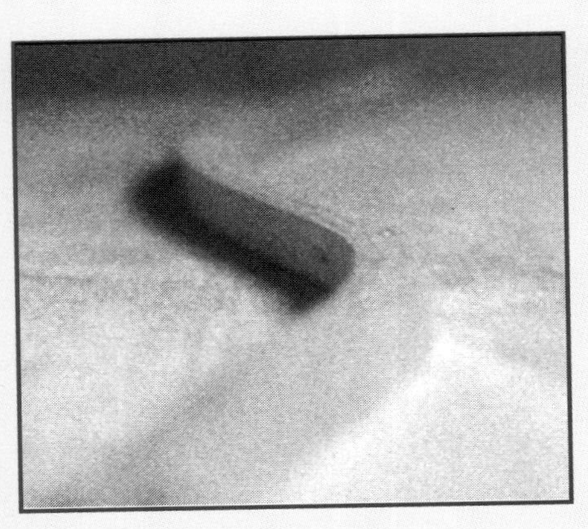

By sighting down the batten from the end, it's obvious if the curve has any lumps or hollow spots. When the curve is even and smooth, it's "fair."

filing by hand or sanding the edge to the contour, we're going to fit a router with a flush-cutting bit and trim the plywood pattern to the exact contour laid out with the batten.

Making the cuts

This operation will work the same whether you are cutting the pattern piece or using the pattern itself to cut duplicate parts, but this section will concentrate on cutting the pattern using the battens attached with brads.

The first step is to set up the router. The shank-mounted bearing rides on the batten and the cutter shaves the edge of the plywood pattern below. Set the depth of the cutter so that the bearing has plenty of contact with the batten and the bit makes a full cut on the pattern.

The guide batten guides the bit along the curve. The track batten stabilizes the router by supporting the router base. It's critical that the two battens are far enough apart to support the base and close together enough so that the base will be supported the entire length of the cut.

Start with the router firmly seated on the two battens, clear of the cut. Then enter the cut, moving quickly and smoothly along the plywood to the end, keeping the bearing in contact with the batten the whole time. Make a second pass just to make sure you've not left any small humps.

Most woodworkers tend to move the router too slowly at first. This results in an inferior cut. Push the router along at a fairly brisk pace. Don't force it, but apply a bit more pressure than you might be used to. Remember, you cut very close to the line with the band saw, so you have very little material to remove.

Keep in mind that when cutting the edge of a piece of good plywood, it is very likely you'll dull your carbide bit in the process of making the pattern. As aggravating as this is, the process sure beats doing the work by hand. I keep two bits handy—one for cutting the pattern and the other for cutting the actual material.

Attaching two battens—one for the router bearing to ride on (the lower batten shown here), and another the same size to keep the router upright—makes short work of cutting this curve.

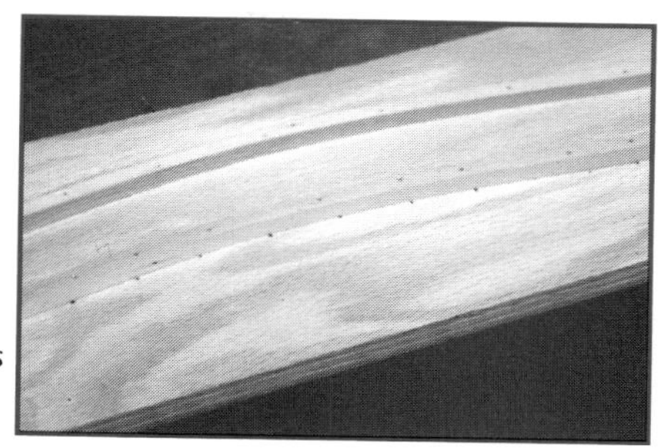

Once you have the batten nailed along the curve, cut as close as you dare with the band saw.

With this setup, curved cuts can be made about as fast as you can walk.

Here is the finished cut. Sometimes when only one side of a piece is going to show and you are making only one of them, you can run the battens on the underside, cut the piece itself just like we did with the pattern, and have a finished curve. This saves the step of creating a pattern.

MAKING A SHANK-MOUNTED ROUTER BIT

You'll need a special flush-cutting router bit to trim the pattern—one with a shank-mounted bearing. These are available from some manufacturers but here is how to make your own.

With the bearing on the shank instead of on the end of the bit, the jig can be used right side up with the work much more visible. More importantly, however, the bearing itself, the part of the machine that actually guides the cut, is very solidly mounted so as to dampen vibration. With the bearing out on the end of a long flush-cutting bit, any oscillation in the bit is magnified. With the bearing closer to the machine, this effect is minimized.

A shank-mounted bearing really comes into its own when cutting very thick (about 2 inches) material. The cut doesn't have to extend all the way through the thickness of the piece, but can be done with two separate passes.

A standard bearing having a 1/2-inch ID has an outside diameter of 1-1/8 inch. You'll need a 1-1/8-inch straight bit and this standard bearing to make your 1/2-inch-shank flush-cutting bit. If you plan to use a 1/4-inch shank, get a 3/4-inch straight bit and use a 1/4-inch ID and 3/4-inch OD bearing. The bearing will slide right on to the shank of the router bit.

The inner race of the bearing should not spin on the shank. Make a small indentation in the shank with a center punch. Support the shank on the bench top and tap the shank with a center punch right where the bearing will ride on the shank. Then tap the bearing onto the shank over the indentation. This will keep the inner race of the bearing from spinning on the shank of the bit. ■

Fasten the bearing to the shank of the bit so the inner race doesn't spin. A prick with a center punch is all it takes.

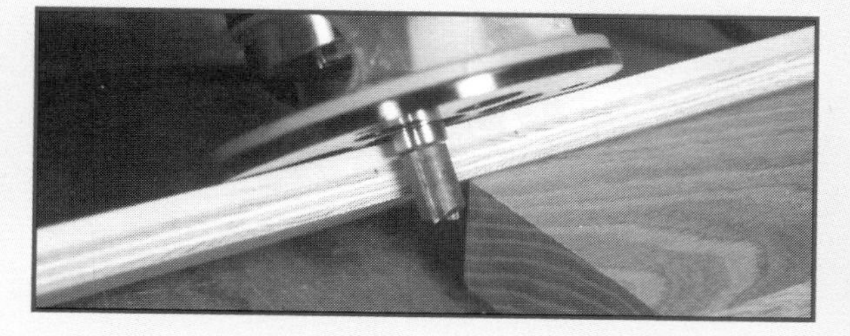

The key to this batten cut is the shank-mounted bearing. This pattern bit, or some variation of it, is a fixture in most professional woodworking shops.

MILLING A 180- OR 90-DEGREE ELBOW

You can do a lot with this shape, from using it on handrail fittings to making door handles.

I had occasion to do a couple of staircases in which the railing was completely round. In the course of building the components, including some rather complicated transition fittings, what emerged was a system for making small curved parts. These parts can be made quite quickly as segments of full circles. The end product, in this case a fully rounded 2-inch diameter, 90-degree elbow on a 1-1/2-inch inside radius, formed the basis of a system of fittings. The process of making the fittings was quite straightforward once I figured out some basic steps. Using this as a guide, the notion of milling shapes in this manner might be useful as a model for other similar shapes.

As you might guess, the starting point was a drawing. On paper, the notion of using full-round 2-inch dowel-shaped railing became much more feasible when I discovered that I could readily purchase pre-milled 2-inch full rounds in lengths up to 16 feet. That left only a myriad of transition pieces to mill and assemble. On the drawing, a 1-1/2-inch inside radius fitting looked like it would be perfect for making the transitions.

How, I wondered, would I be able to hold such a small piece to mill it? The mental breakthrough came at break time in the form of a glazed donut. Suddenly, the notion of cutting 90-degree pieces out of a donut made perfect sense. That's exactly what I did.

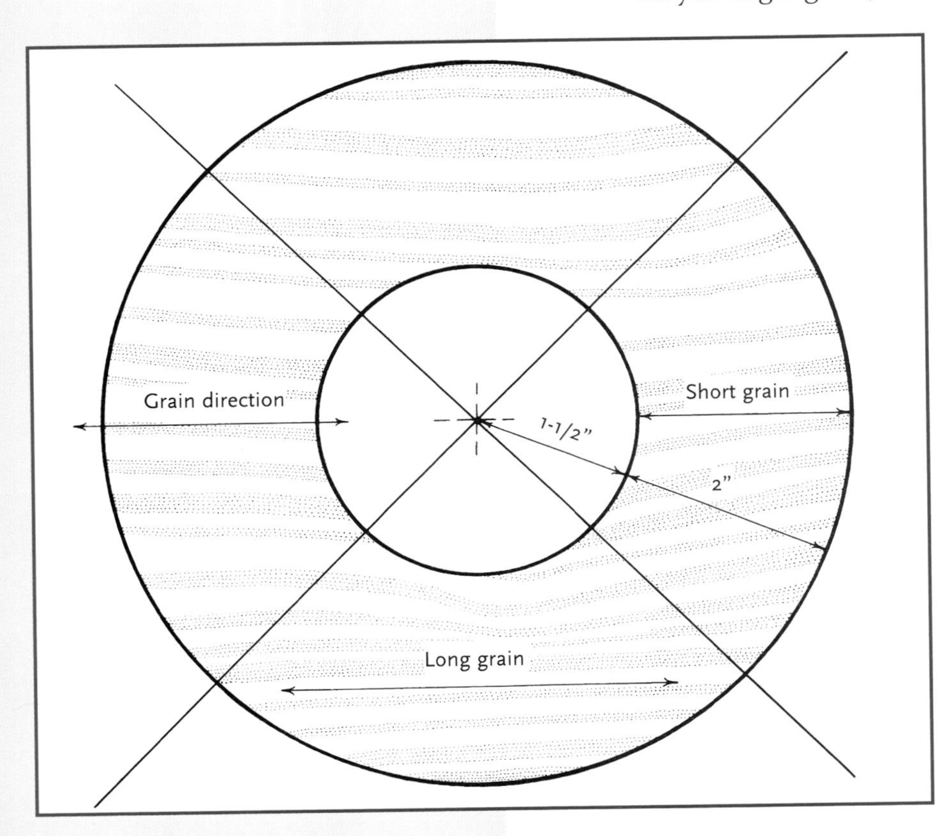

It's best to cut only two elbows out of a full circle. This ensures that the grain will remain structurally sound.

The next items to tackle were how to go about making the donuts, material requirements, and a sequence for producing the 90-degree elbows.

With the grain orientation a vitally important part of the mix, I determined that I would cut only two elbows out of each full circle. Why? I wanted the grain to be as structurally sound as possible. This, after all, was hand railing. It had to be strongly built. By discarding all but two elbows in a full circle, I could cut two elbows with the grain direction at its strongest.

Cutting the donut

Please make special note of the fact that this project is not for a beginning woodworker. The process of making a solid block into small parts is something you should attempt only if you have a fair amount of experience on a router table.

To mill a 2-inch donut with an inside radius of 1-1/2 inches, I would need a block 2 inches thick and a little over 7 inches square. This was simple to glue up out of 5/4- or 1-1/16-inch stock. I glued up enough stock to cut out all of the squares I'd need for the entire project. I planed the wood to exactly 2 inches in thickness, then I drew circles where the cuts for each donut would be, allowing plenty of space between them.

Next, since I wanted a 1-1/2-inch inside radius, I purchased a rather intimidating looking 3-inch diameter drill bit. I drilled 3-inch holes into the wood at the center of each donut. In this way I was able to produce a perfect inside cut. It is important to note how dangerous a drill press can be if the part being drilled spins all of a sudden. Be sure to use clamps to hold the part down to the drill-press table.

I used the band saw to cut the outside diameter of the pieces. A quick sand to even out the circumference of the pieces and it was off to the router table.

Making a donut

It takes a block this size to make a single donut.

Rough out the outside circle with a band saw.

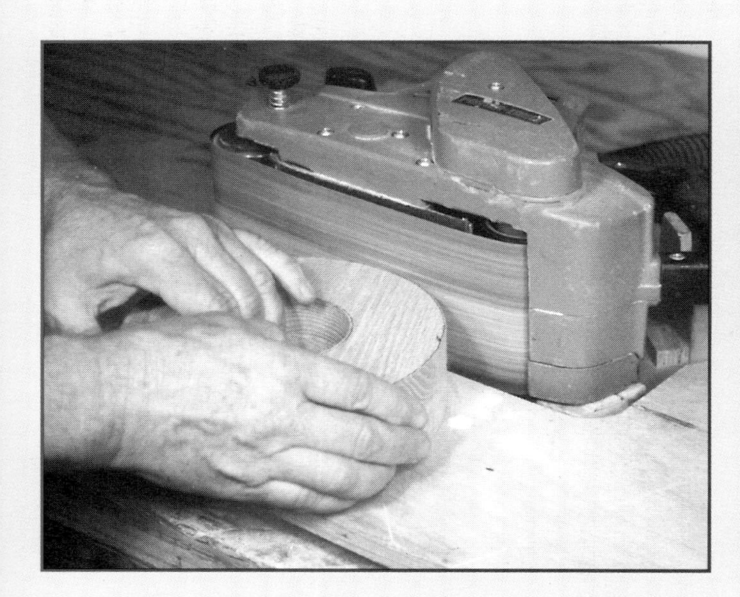

Smooth the outside with a sander.

Using a 3-inch diameter drill bit, the interior 1-1/2-inch radius is easily cut.

Shaping the donuts

Again, I would caution that this step is not for beginners. At this point the pieces are rounded inside and out, 2 inches wide and exactly 2 inches thick. The next step is to round the edges. Since the sectional dimension is 2 inches, a 1-inch radius on each edge works fine—a 1-inch radius round-over bit makes short work of this step.

A 1-inch radius round-over bit produces the 2-inch thick full-round shape of the donut when run on all four edges.

Using this bit in a router-table setup seemed much safer to me than trying to secure a cylindrical donut while using a portable router. In a way, the donut makes its own holding jig. But this is one time where having a very clear idea of the direction of the bit rotation is vitally important. If you haven't already done so, I recommend that you take the time to draw an arrow on your router table so that you know for sure which way the bit spins.

By way of warming up to the shaping process, I make very light cuts, usually at least four or five passes before I put the bit at full height. The beauty of this kind of cut with a bearing-mounted round-over bit in a router table is that gradual cuts are straightforward to make—simply raise the bit gradually with each successive cut. If you have several pieces to cut, make all four of the passes on all of the pieces before you raise the bit on the router.

Initially, the bit protrudes from the table only a small amount, taking only the lightest of cuts. The bit is raised about 1/8 inch for each subsequent pass until it's at full height. Multiple passes give you lots of practice for that all important final pass.

This is a good time to discuss the process of handling this type of router cut. Start the cut in an area where the grain is in line with the cut rotation. When cutting a full circle, you'll be cutting end grain, "'downhill" grain, and "uphill" grain as you rotate the piece. Start the cut "downhill." (See the drawings on the following page.)

The terms "uphill" and "downhill" refer to the action of the cutter in relation to the grain of the wood. Visualize using a spokeshave to shape this part. It's impossible to get a good cut with an edge tool going against the end grain, or "uphill." You would have to turn the tool around and shave "downhill." It works the same with a router, at least to start.

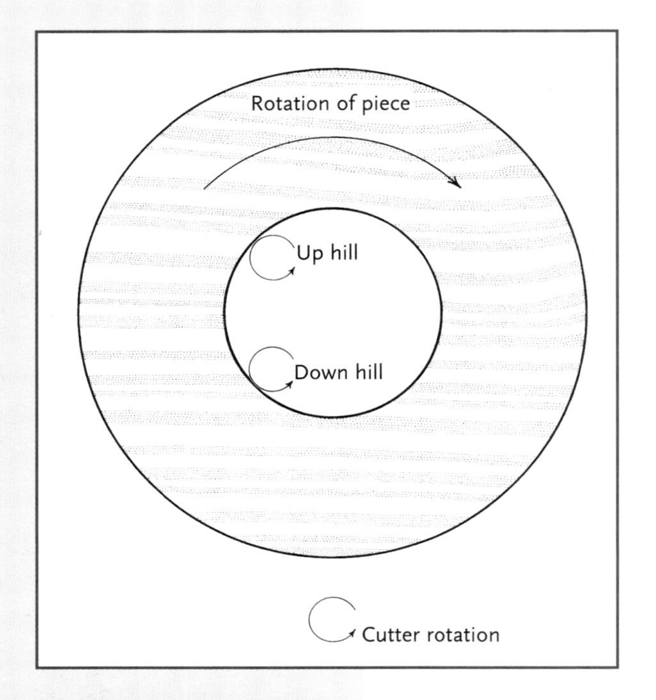

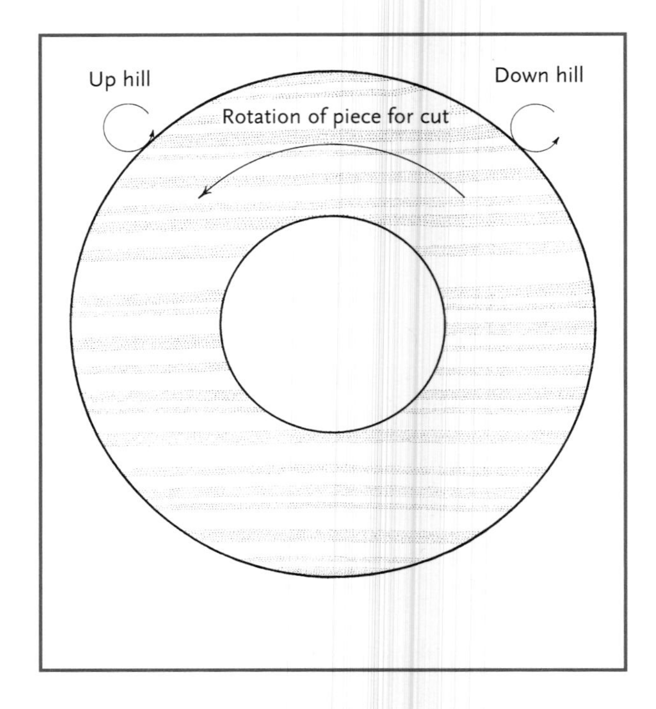

"Uphill" and "downhill" relate to the action of the cutter in relation to the wood grain.

With a rotating cutter, you don't have this option when cutting a circle. There will be places where the cut is directly uphill. The result will be a mediocre cut under the best of circumstances. It's simply a fact of life. The best you can do is to make sure you start in a spot that's downhill and move through the uphill portions smoothly—another good reason for taking light cuts. The more you practice the cuts, the better you get.

Routing the outside circumference

It's tempting to think that the entire cut must be made by rotating the piece against the bearing without stopping. In fact, by rotating the donut, it's almost certain that burn spots will appear as you shift your grip on the piece. A better strategy is to grip the piece, make a partial pass on the bearing, and then pull the piece away from the cut. Then grip the piece in a new spot, start the bearing on an area previously cut (overlapping the cut in a way), then make another pass on part of the circumference. Draw the piece away again when you reach the limit of your grip on the piece.

This method has several advantages. First and most importantly, your hands never leave the secure position you begin the cut with. There is no fumbling with the piece while it's in contact with the cutter. Secondly, by overlapping the cut and making a firm pass on just part of the circumference, the cut quality is much better, with no burn marks where you had to hesitate. Thirdly, by making very light passes, by the time you get to the final pass you've built a rhythm that helps you produce a high-quality cut so there's much less sanding to do.

By cutting a full circle to get the two 90-degree transitions, the cuts are simple and the shape acts as its own holding jig.

Routing the donut hole

When you're routing the inside of the circle, the idea is to carefully place the donut over the spinning cutter and run the bearing around the inside of the circle. This is another time when knowing the rotation of the cutter is important, and when making light cuts to get used to the feel of the machine really pays off.

If you look down through the center of the donut as you place it on the router table, it is quite straightforward to get it in the right place. Practice a couple of times with the machine stopped to get used to the idea. Once the piece is in place, free of the cutter, begin the cut in an area where the grain goes in the direction of the cut—"downhill." This will be the smoothest area in which to start the cut.

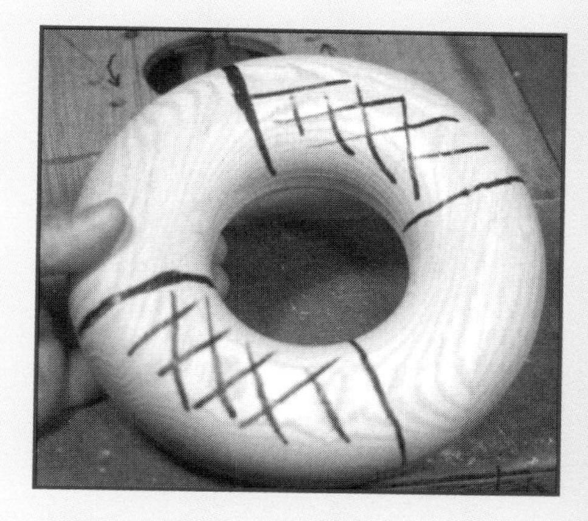

Only two 90-degree elbows can be cut from a full circle if the grain is going to be diagonal.

The finished elbow

Once you've got the donuts fully rounded, you can cut out the two 90-degree elbows. Rough-cut them on the band saw, then trim them in a miter saw. The result is a fitting with the perfect grain orientation, a radius that's just right, and a full-round shape made in a jiffy. Yes, the process does use a great deal of material to produce a rather small part, which illustrates the main downside of milling curved parts.

PART THREE

Lamination-Bending

Lamination-Bending

THIN LAYERS OF WOOD, when glued and clamped over a curved form, are a reliable way to make curved parts quickly and accurately. The process is called lamination-bending. Lamination-bending can be tricky at first, but, among other things, this technique allows the use of woods that don't steam-bend well (as discussed in Part 4), and it solves the structural problems that are inherent in milling a curved piece out of a solid board.

A whole series of variables has to be controlled for bent lamination to work reliably. Each variable, though, from the type of glue to the thickness of the laminates, is easy to understand. You can expect consistent, high-quality results right away when you use the right setup, materials, and techniques. Despite being messy, gluing up the laminations is easier than it might look, as long as you have enough clamps.

Which brings up an old woodworking riddle: How many clamps do you need in a woodworking shop? The answer: One more than you have. At last count, I had 175 clamps of various sizes and configurations. Although new developments like vacuum-bag clamping are changing equipment needs, sometimes I still could use that one extra clamp.

ADVANTAGES OF BENT LAMINATION

Thin layers of wood properly glued together form a construction that is as strong as or stronger than a solid or steam-bent board. Modern glues bond with a strength exceeding that of the material itself. Unlike in milling, where the strength of the piece diminishes rapidly when the degree of arc of a curve makes the grain go diagonally across the board instead of along its length, bent lamination has no such limitations. A full circle is possible with no end grain or short grain problems.

Even exotic woods that typically don't steam-bend well can be curved if they are cut and laminated. If the wood you would like to use steam-bends poorly, such as mahogany, for example, bent lamination is a good alternative (see Part 4).

Once the initial springback takes place, the bent piece is strong and stable. It will have little or no tendency to move, creep, straighten, or otherwise become unstable. This assumes, of course, that you've used the correct methods, glue, and materials.

Once you perfect the setup, materials, and techniques, it will be entirely possible to produce identical parts even on a production basis. Suppose you were to make a series of rocking chairs: Using bent lamination for the runners would be a great way to help you consistently get the shape you want.

With bent lamination, there is relatively little waste, especially if sliced veneers are used for laminates. By comparison with milling curved parts or steam-bending them, the waste factor is low. If there is a flaw in a board that would make it unsuitable for other methods of making a curve, it may often still be used as a bending laminate. Furthermore, scarf joints (pg. 102), allow you to connect short laminates and make pieces of virtually any length. Narrow boards can be edge-glued to make up wider ones.

BENT-LAMINATION RULES OF THUMB

Thinner laminates are preferable to thicker laminates.

It's easiest to glue up all the laminations at one time.

Make the bending form as perfectly curved as possible.

Don't use white or yellow glue.

Make a dry run before you mix the glue.

Work quickly when gluing and clamping.

Allow for springback. ■

DISADVANTAGES OF BENT LAMINATION

Glue—and the gluing process—is at the heart of many of the disadvantages of bent lamination. Safe, flexible glues aren't suitable, and the less flexible glues like plastic resin and epoxy can cause severe reactions in certain individuals. The user must protect his or her skin, eyes, and lungs from overexposure not only during mixing and clamping, but during sanding as well. Gloves, masks, and long sleeves are must-haves, and a good vacuum system must be in place to minimize exposure to dust.

Aside from the inherent hazards of the glue, the process itself can be quite messy. Excess glue will get on clothes, benches, and floors, and it must be cleaned up constantly. And many glues appropriate to lamination-bending do not wash out of clothing.

Cost is another factor. Not only is the glue expensive, but the process is slow since the glue requires extensive setup time. After setup, the work still requires much sanding and shaping. In addition, a great deal of material is used when strips are sawn from solid stock for bending. Cutting the strips in the shop represents an additional step. There is added expense if veneers are purchased, although when this is weighed against the savings in labor of cutting strips yourself, it's often a wash. Other advantages to using veneers, such as the ability to match the wood grain, help offset this cost.

Take away all of the other disadvantages and you are still left with the unavoidable glue lines in the finished piece. Even when the process is done very well indeed, these lines are plainly visible. This problem grows worse in lighter-colored woods when darker-colored glue is used, for example, when plastic-resin glue is used with maple or ash.

Here's the problem: When making a round section out of a bent lamination, toward the top and bottom of the section, the glue line is cut at a very steep angle. What this does is create a very wide glue line. Sometimes this doesn't matter, but when it does, the glue line can come as a rude shock as soon as you apply stain.

Most furniture I see these days (hoop-back chairs, for instance), uses a square section with slightly eased edges instead of a full round. One reason for doing this is to avoid the problem of wide glue lines.

Having said this, it is entirely possible to get a nice looking bent lamination. The example in the photo at right is about as good as I've ever managed. The board was first ripped into strips. By keeping the strips in the same sequence as they were in the solid board, the grain pattern looks natural. Effective clamping makes the glue lines fairly narrow. By using thicker outer layers (photo, lower right), otherwise noticeable glue lines are virtually eliminated. Can you see the glue lines? Yes. But you have to look really closely.

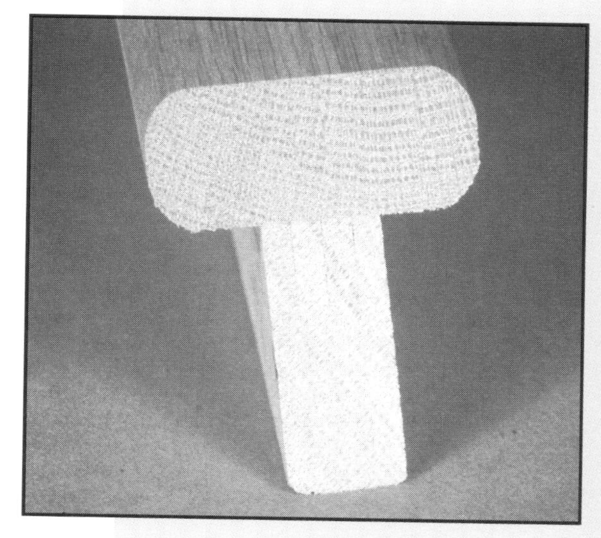

Seen from the front, the top piece of the tee looks solid, with only slightly discernible glue lines.

PRINCIPLES OF BENT LAMINATION

How does lamination-bending work? What keeps the piece curved after the glue cures? If you take a paperback book and bend it into a U-shape, you see that the pages slide on one another. This is what allows the book to bend. If you were to spread glue between each of the pages before bending the book this way, when the glue dried, you would wind up with a permanently curved book because the glue would prevent the pages from sliding back. A lamination bend stays bent because the glue, once cured, prevents the pieces from sliding back on one another. Period. There is nothing but the glue keeping the pieces from straightening out.

Seen from the rear, however, you can tell that the top piece is made from nine pieces glued together into a lamination. For the seamless look, you have to cut the layers out of a solid board, then make sure you glue them back together in the same order as they were cut.

The bend remains under constant stress even years after the glue has cured. If you could dissolve the glue from a circular handrail bent 20 years previously, you would see the pieces straighten out as if they were bent yesterday, retaining only slight curvature. Always and forever, the wood would prefer to be straight again. What this means is that the glue joint is under constant pressure. Only glue designed with pressure-resistance in mind will stand up over time. White and yellow glues are not designed for such an application because they remain flexible even when fully cured.

DETERMINING THE THICKNESS OF THE LAYERS

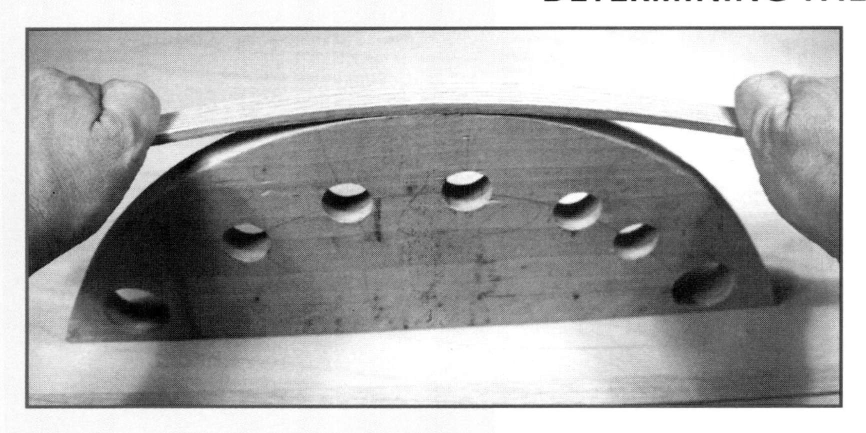

If a laminate doesn't bend readily around the radius, it means the entire bundle won't bend very well. The layers each need to be thinner.

The generally accepted practice is to mill the laminates as thick as possible but as thin as need be. Usually, the thinner the better. Thin laminates put less stress on the glue joints, have less springback, and make a stronger construction. It's therefore better to err on the side of thinness rather than thickness. There are many ways to find the thickness of the laminates, including math formulas, but they should be used as a starting point, no more.

A low-tech, nearly foolproof method for finding the correct thickness for an individual laminate is to cut a single layer of the material you plan to use. Try to bend it around the radius you're working with. If the test piece feels so stiff that it might crack, it's too thick. It's that simple.

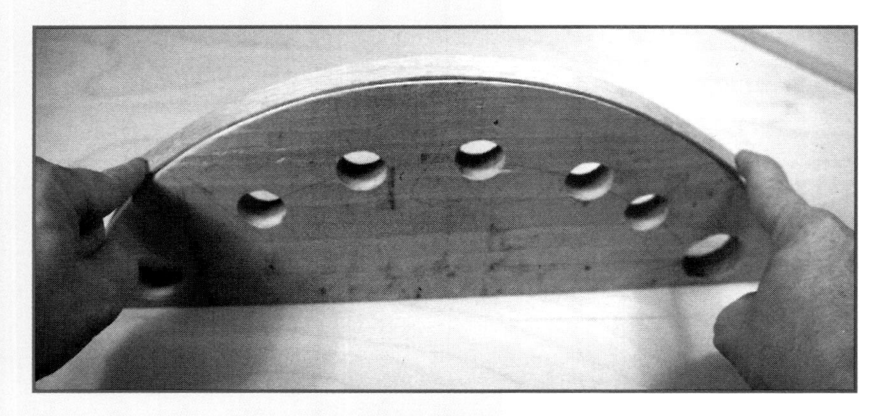

The thinner the laminate the better, but at the very least the laminate must be able to bend easily around the form.

Mill the laminate thinner and thinner until it bends around the form quite easily. By "quite easily" I mean that the piece should be flexible enough to be held on the form with no more than two fingers. If the piece makes a popping noise like the sound of cereal when milk is poured on it, the wood is telling you it's too thick. Each individual laminate must be thin enough to bend easily around the radius you're using. Hence, the smaller the radius is, the thinner the layers will need to be, and the more of them you'll need to build up to the finished size.

Once you've established the correct thickness of the layers, mill enough to stack to the overall size needed plus a little more—an extra piece or two will make the bundle the right size. Do the same when cutting the laminates to width and length. You'll be sanding and shaping after the glue sets so you'll want extra material to work with.

Soaking the laminates

While soaking the layers in water may seem to make them more flexible, it actually does more harm than good. The act of soaking leaves a film on the gluing surfaces that partially prevents the glue from penetrating. In addition, there is no good way to tell if the wood is really dry, another crucial factor in gluing. Successful gluing is difficult enough without extra variables. You can avoid soaking by simply making the layers thin enough to bend easily. Soaking is simply one additional variable you don't need.

COPING WITH SPRINGBACK

The subject of springback goes back to the way lamination-bending works in the first place. The layers slide past one another during the bend. They try to slide back again when the clamps are released.

When the clamps are released, the glued pieces will relax and try to straighten out a bit. Just how much the piece will spring back is determined by a number of factors, the most important of which is the total number of laminates in the construction. The good news is that once the piece initially relaxes, assuming you've used the right glue, it will not straighten out any more.

SPRINGBACK RULES OF THUMB

The thinner the laminates, the less springback there will be.

More layers mean less springback.

The larger the radius, the less springback.

The smaller the degree of arc (distance around the circle), the less springback.

The more flexible the material, the less springback—poplar is more flexible than maple, for example. ■

SCARF JOINTS

Many times over the years I've needed a board or plywood piece longer than I could find. Most hardwoods are cut to 16 feet and no longer. Dozens of times as a stairbuilder I've needed to make circular handrails longer than 20 feet. What to do? The scarf joint is the answer.

Scarfing is a way of joining two boards, or two sheets of plywood, end to end by overlapping long, tapered glue joints. Properly done, the scarf joint is nearly invisible and just as strong as the board itself. Using this technique, it's entirely possible to make boards and plywood panels as long as you like. Where bending is concerned, the beauty of making a joint that's just as strong as the board is that the joint retains the same flexibility as the rest of the board. The key to making the joint is to get the slope angle just right.

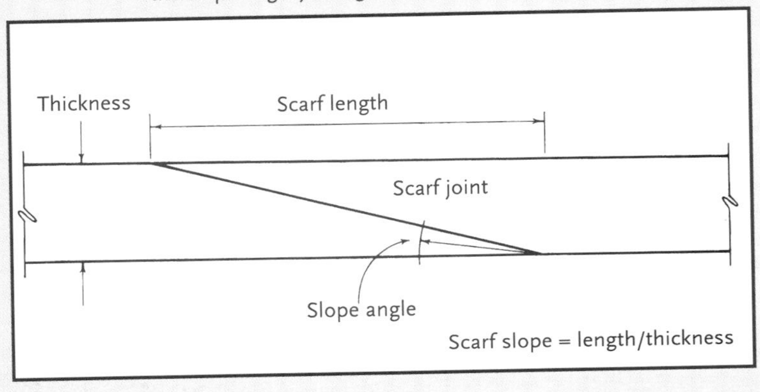

A scarf joint is a long bevel cut on the ends of two boards and glued together to lengthen the two into a single long board. Carefully done, the joint is nearly invisible and as strong and flexible as the board itself. By scarfing, boards of almost any length can be made up of shorter segments. The slope of the scarf shown is 4:1.

Slope angle

Boat builders, the most frequent users of scarf joints, differ on the perfect slope, but the consensus is somewhere between 8-to-1 and 12-to 1. If the slope were 8-to-1, a board 1 inch thick would have an angled cut starting with the full thickness of the board and tapering to a feather edge over 8 inches of length; 10-to-1 tapers over 10 inches, and so on.

Experiment with different slope angles until you find what works best for your particular situation. The longer the joint, the greater the amount of material on each piece you'll need to lengthen your board. The trade-off is, the longer joint is stronger and more flexible for bending.

Glue up scraps then bend them until they break. Try to break the piece right at the scarf joint. You'll probably soon discover that the amazing strength and flexibility of a scarf joint largely depends on the slope angle.

Cutting a scarf joint

I've tried every tool and technique imaginable, from a power plane to a band saw to a hand plane to a router. The trouble with all of them is that the feather edge at the end of the board gets quite fragile as you cut it. By all means try all of these tools and any others you are inclined to use. But every tool I've tried tears out the knife edge of the bevel, except one. I finally tried a belt sander. I've never looked back. To cut scarf joints quickly and accurately, all you need is a good belt sander with a new 100-grit belt, a clamp, and a tape measure.

Setting up the bevel cut

Using 1/4-inch thick material for the layers, a pretty typical size, I simply stack the layers on the bench like steps, then sand off the steps to form the slope. I put an extra layer under the stack and another on the top to make the pattern easier to follow. That's really all there is to it.

If we use an 8-to-1 slope and we want to finish at 1 inch thick, all we need do with 1/4-inch thick material is expose 2 inches of each layer to form the 8-to-1 ramp. Carefully measure and stack the layers so that each has 2 inches exposed, clamp the stack to the bench, and sand smooth. When you have sanded the exposed ends to a feather edge, you will have shaped the scarf joint perfectly.

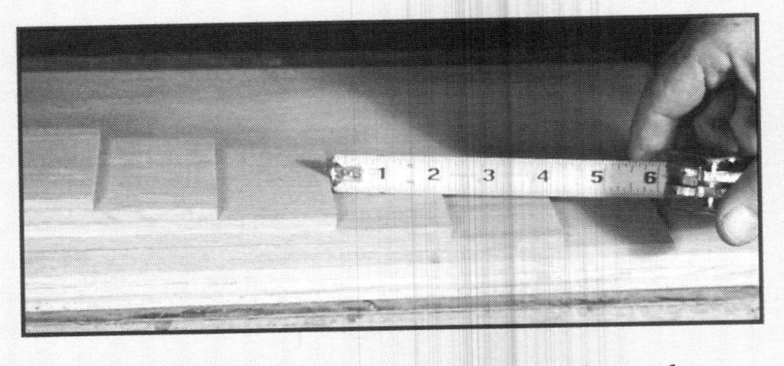

To cut the scarf joint quickly and accurately, stack up the layers like steps, then sand the stack to form a smooth ramp. The slope angle is determined by the exposure of each step, in this case 2 inches for each of the 1/4-inch thick layers, resulting in an 8-to-1 slope angle.

Belt Sander

I am going to assume a certain degree of expertise with a belt sander, but it doesn't take an expert as long as you follow a few guidelines. Don't just grind away endlessly without looking at the result. Stop every few seconds to see where you're actually sanding and where you're not. Concentrate on getting as clean a bevel as you can. Make sure the cuts stay even as you sand.

Watch the lines of the exposed ends as you sand. They should stay straight and square. If they become curved, or if the top of the stack starts to look different from the bottom of the stack, selectively sand until you get back to a uniform pattern.

A belt sander makes short work of removing the steps to form a ramp. Notice the layers above the sander—the sanded edge is not square. To correct it, simply sand on the high area until the angle is square again. To maintain uniformity, be sure to sand the entire length of the ramp equally.

Gluing

I use either epoxy or plastic resin glue for gluing scarf joints together because the joint will be under considerable stress when it's bent as part of a bent lamination. Depending on the slope, the scarf will be more or less end grain and so the wood may soak up extra glue. Coat the parts and place the joints together, but don't yet clamp. Let the glue soak into the wood for a few minutes. Then open the joint and coat the surfaces with more glue before clamping.

When I'm scarfing a number of thick pieces for a bent lamination, I like to glue up several of the joints at once. As long as the material is fairly small, say 1/4 inch thick by 2 or 3 inches wide, it's easier and faster than gluing them up one at a time. Just place plastic sheeting or pieces of waxed paper between the layers to keep them from sticking together.

It's important to set up the gluing operation to maintain consistent thickness on the board and also to make sure it's straight. If the parts overlap too far, the resulting joint will be too thick. Conversely, if the parts don't overlap far enough, the joint will be thinner than the adjacent material. I usually overlap the parts just a little extra, to make a slightly thicker zone at the joint. Make sure the pieces are in alignment either by sighting down the length of the glued pieces or by pulling a string tight near the edge of the boards as a guide. Then, after the glue has cured, sand or plane the board to the proper thickness, measuring with a dial caliper to make sure you have it just right.

If the parts are overlapped too far, the joint will be too thick. If they aren't overlapped enough, it will be too thin. Make it just a bit too thick, then sand or plane the joint to the correct thickness after the glue cures. Using this technique it's entirely possible to make a board that's 50 feet long.

If you follow the guidelines for figuring proper laminate thickness, springback will be but a minor factor in determining how to plan the bend. The smaller the radius is, the thinner the layers will need to be to bend easily around the radius. The thinner the layers are, the more of them you'll need to build up to the overall size. The more layers you have and the thinner they are, the less springback you'll need to allow for.

How much springback should you build into the form? Many have tried to provide a scientific answer to this question, but in the end, good old trial and error prevails.

All bent laminations spring back somewhat. Let's consider two extreme examples. The first example has a 10-foot radius, using sixteen 1/16-inch thick veneers bundled to a height of 1 inch. The second example has a 10-inch radius. It is the same height, but uses only four laminates that measure 1/4 inch thick. In the first example, we use thin laminates over a gradual curve. In the second example, we push the limits of the material by using a thickness that nearly reaches the breaking point in bending around the much smaller radius. The many layers of laminates and gradual radius in the first example result in springback that is hardly measurable. In the second example, however, when the clamps were released you would see noticeable springback.

This piece has no springback whatsoever. The layers are quite thin for the radius and there are lots of layers, both keys to minimizing springback.

THE BENDING FORM

The forms for lamination-bending and steam-bending are similar. Both types of forms must be fair, strong, and provide for fast clamping. Remember that any lumps or bumps in the form will appear in the bent lamination, so the curvature of the form must be as close to perfect as possible.

Male bending forms allow gradual bending and clamping. To use a male form, you begin clamping either in the middle or at an end. All forms for bending must be smooth and allow for fast clamping. (Clamp pockets, a means to clamp the form securely, make the process go more smoothly.) Layers of plywood make a good form.

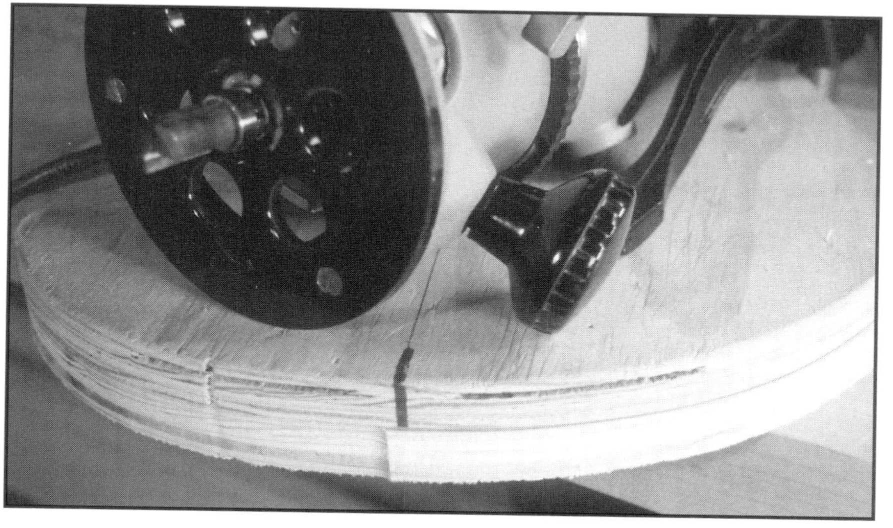

By using the shank-mounted bearing flush cutter, you can duplicate as many layers of form material as you need, matching the contour exactly.

Forms are typically made from MDF or plywood. It works best if the form is the same thickness as the bundle of laminates, so building up layers of plywood or MDF is the simple solution. Start with a fairly thin (1/4-inch) piece and shape it as smooth as possible. Using this piece as a pattern, flush-trim enough plywood panels to make the thickness you need (see the section on patternmaking, pg. 84). The form must allow for springback so that when the piece is unclamped, it straightens out the predicted amount. So the glued part doesn't stick to the form, use wide cellophane tape on the surface of the form to act as a mold release.

Female forms rarely work for bending. Getting the pieces into the curved shape can't be done gradually, as it's possible to do with a male form.

Types of forms

A male form has real advantages over a female form or a two-part form. Bending a bundle of laminates around a male form, clamping as you go, pulls the piece into the form gradually—a real plus. Remember that pieces slide past one another as they bend—with a male form, you can start at one end or in the middle and work outward, which allows laminates to slide as they're secured.

By comparison, a female form distorts the bending process. The ends touch the form first, then the middle of the bend is pulled into place. This has the unfortunate effect of concentrating the clamping and bending pressure in three spots: at the ends, where the bundle of laminates first touches the form, and right at the center, where the clamp is. This can distort the bend beyond the breaking point.

For those times when an S-shaped bend is needed, a female form may seem logical, but a better solution is to use two male forms, as shown on pg. 108. This way the bend can be accomplished by pulling the free ends of the piece down onto the form.

Two-part forms made from a male and female form clamped together (pg. 109), may seem like a logical way to save on clamps and an equally logical way to evenly distribute clamping pressure. But unless the forms are made extremely carefully, the results probably won't be satisfactory. Remember that the thickness of the bundle of laminates must be cut out of the form, not just a single kerf. For example, if the inside radius for the male form is 24 inches and the bundle is 1 inch thick, the outside radius for the female form must be 25 inches.

That said, even if precisely made to the required radii, two-part forms don't distribute clamp pressure very well. When the two forms are clamped together with the bundle of laminates between, the pressure is concentrated at the crown of the curve and there is practically no pressure at the ends.

A better alternative is a male form with segmented female clamping cauls, as shown below. The blocks apply clamp pressure exactly where it's needed. Clamp pockets cut where they will facilitate the right direction for the clamp make the process go quickly.

Making a segmented female form amounts to nothing more than cutting the one-piece female form into segments on the band saw. Allow distance between each segment so the segments don't bind. Cut clamp pockets and you're ready to go. As with other bent laminations, start at the crown of the curve and work your way toward the ends, alternating back and forth as you apply clamps.

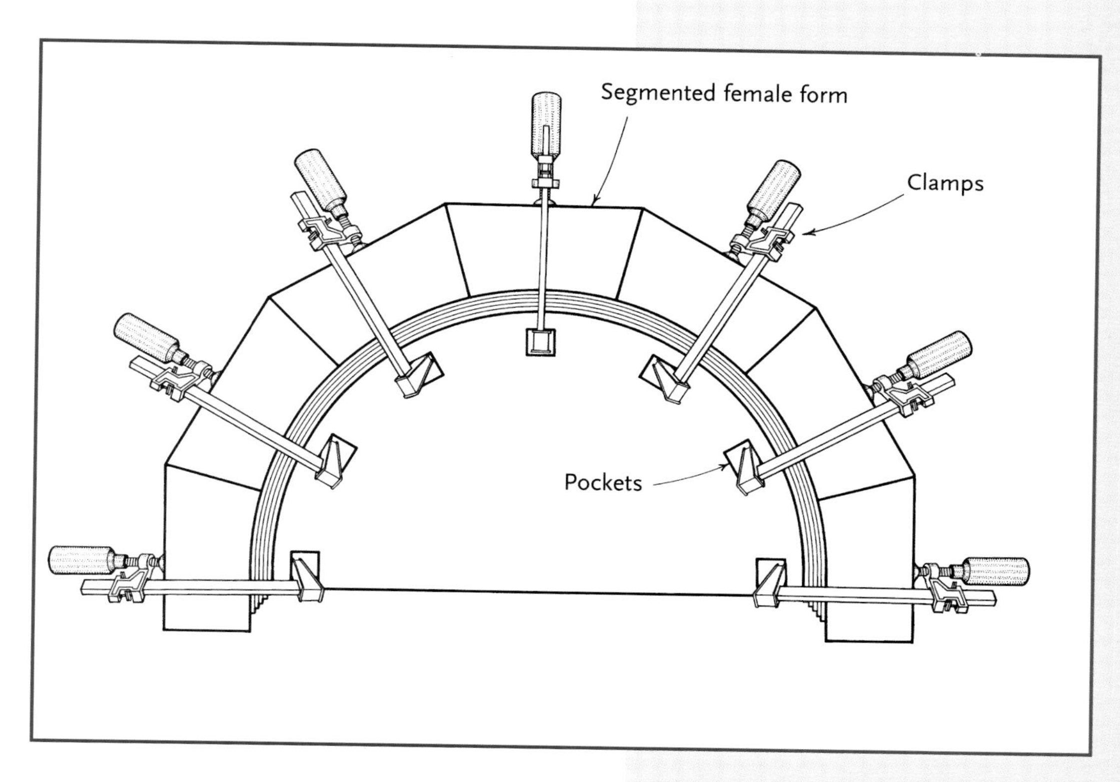

Using segmented female blocks along with a male form instead of a one-piece female form ensures an equal distribution of clamping pressure.

Making an S-shaped bend

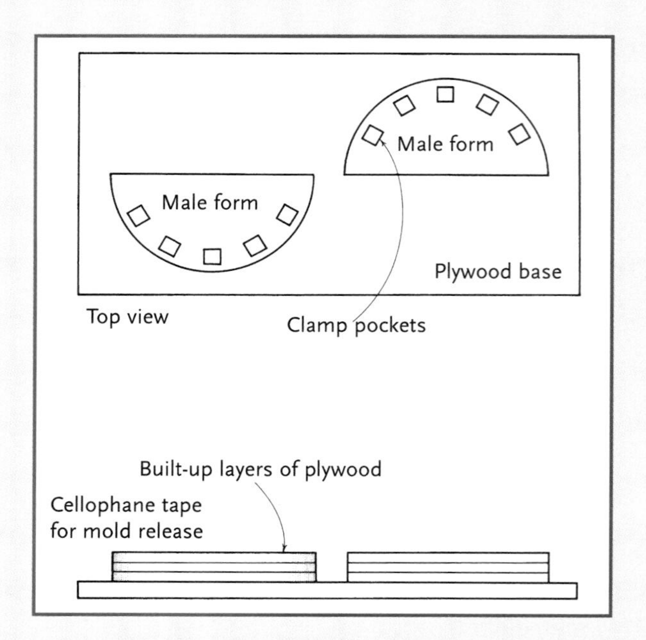

The way to make a form for an S-shape is to use two male forms. Cut them out of plywood, stacking the layers as tall as the width of the piece you're about to bend. Cut lots of clamp pockets. Screw the two male forms to a plywood base. Cover the form with cellophane packing tape to keep the part being bent from sticking to the form.

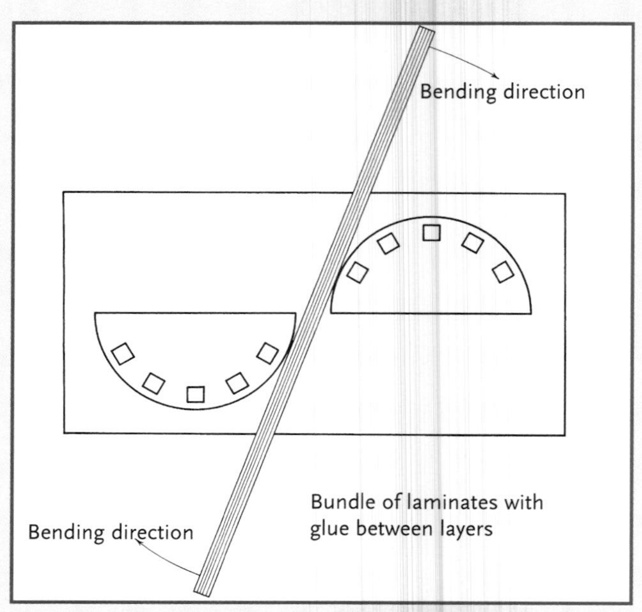

Insert the laminate bundle as shown above. Note the direction in which you'll be bending the bundle.

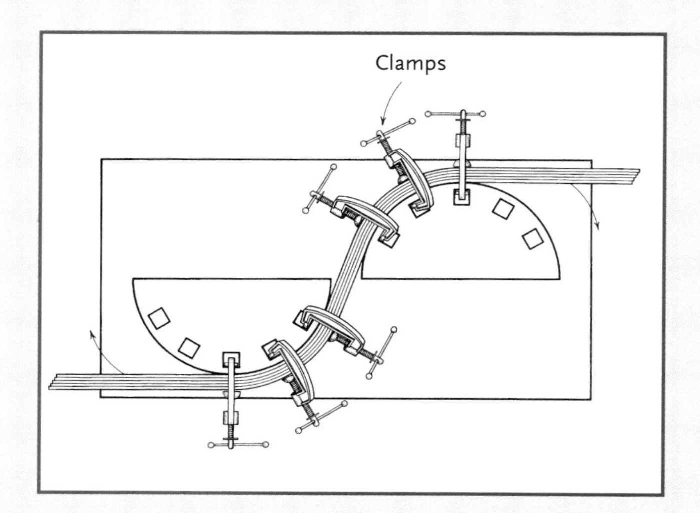

Begin clamping in the middle and work outward to the ends. Clamp as you go. You can see here why you need to cut a lot of clamp pockets when you make a form for this kind of bend.

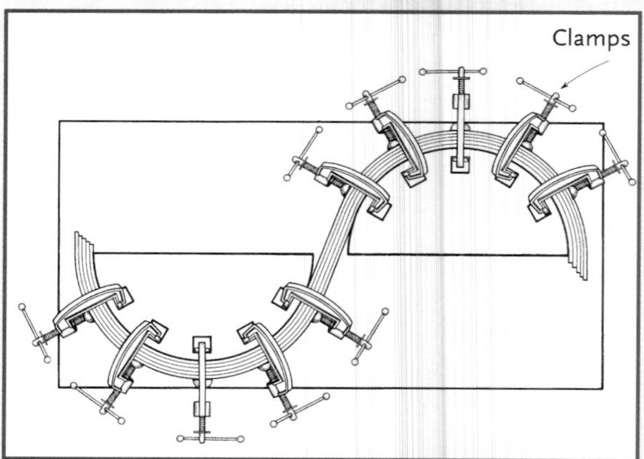

Finish the bend by clamping the ends of the laminate bundle over the forms.

Two-part male and female forms

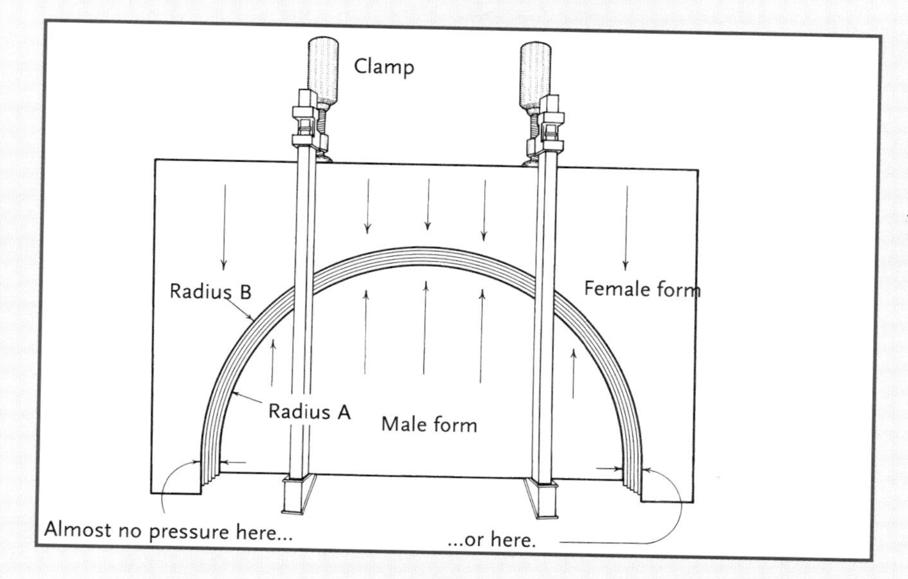

Allow for the thickness of the laminate bundle by cutting different radii for the male and female parts of the form. Still, even if the halves of a two-part form are carefully made, almost no pressure will be applied to the ends of the bend.

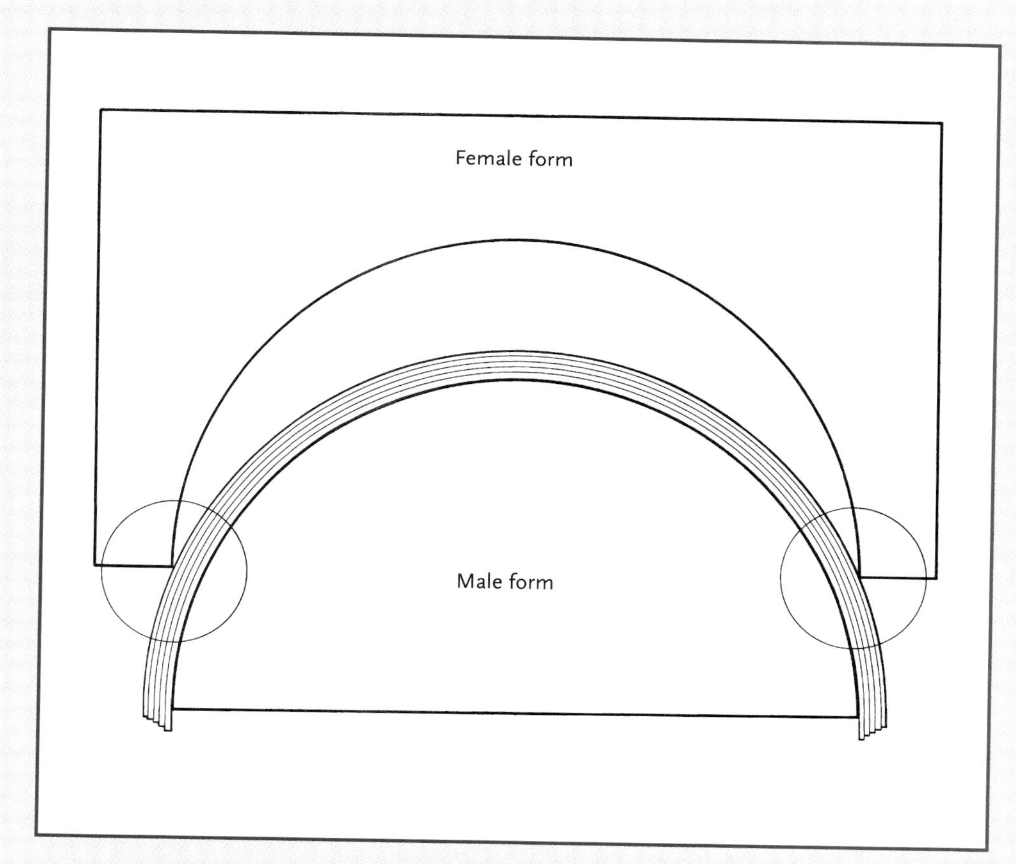

A two-part male/female form won't leave room for the bundle of laminates if you use only a single saw kerf to cut the form pieces apart.

CLAMPS FOR LAMINATION-BENDING

More important than using any one type of clamp is to use a lot of them. An equally important consideration is the amount of pressure the clamp offers. (This will depend on the glue: Some glues, like epoxy, need little pressure. Others, like plastic resin, need more pressure.) The most important issue of all in clamping is the time it takes to get the clamps in position. Very often the single most critical factor separating success from failure is the length of time it takes to clamp the lamination in place. If too much time elapses and the glue sets up before you get your piece clamped, the result could well be complete failure of the bend.

Adjustable bar clamps may seem expensive initially, but become reasonable when you factor in all the use you'll get out of them. They almost never wear out, there are many sizes and configurations available, they supply lots of pressure, and are fast to apply. Good bar clamps last a lifetime and can be used for many things around the shop besides clamping up lamination bends. They are my first choice.

Other common workshop clamps are less efficient. Spring clamps, which are like big clothespins, are fast to apply, but offer too little clamping pressure for all but epoxy-glued laminations. Hand screws and C-clamps are really too slow for gluing a complicated bent lamination. They provide good pressure, and are relatively inexpensive, but unless you carefully set up your operation to accommodate the time you'll need to secure them, their operation slows the process too much.

Shop-built clamps

Shop-built clamps are a good alternative for lamination-bending when the cost of acquiring enough commercially made clamps is prohibitive. They can be made in unlimited quantities from scrap wood, threaded rod, and nuts and washers.

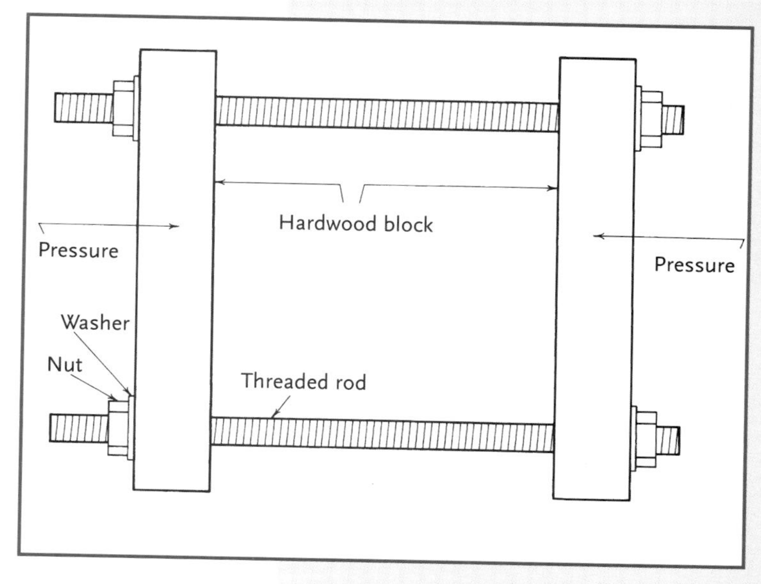

Boatbuilders make bolted parallel jaw clamps like these by the hundreds to glue up mast sections. They couldn't be easier to construct, but a drawback is that they're usually slow to apply.

Spar clamps—Start with some hardwood blocks as big as you need them—1 inch square by 6 inches long is a good starting point. At the home center, get some 1/4- or 3/8-inch threaded rod and ample nuts and washers. Drill the blocks to accept the threaded rod, cut the rod to length, screw on the nuts with washers, and you're ready to go.

Two considerations: First, when you cut the rod to length, run a die nut over the cut end and slightly round the end on the grinder. The last thing you want is to fuss with starting nuts when you're in a hurry. Second, when applying the clamps, use a long socket with a cordless drill to apply pressure. It'll save lots of time.

PVC clamps—When less pressure is needed, as when gluing with epoxy, sections of PVC pipe are perfect for clamping. The process of making these clamps involves slicing off sections of pipe then cutting out part of the section. The C-shaped clamp is stretched open, and clamps down on the part when it's released. The beauty of these clamps is that not only are they fast to make, they're fast to apply. The clamping pressure of PVC clamps is controlled by varying the pipe diameter, wall thickness, width of clamp, and size of the cutout. For example, a 2-inch wide section supplies more clamping pressure than a 1-inch wide section. A small cutout supplies more pressure than a large cutout. A bit of experimentation will soon clue you in to the ideal sizes for your application.

The pressure of PVC clamps can be varied by changing the wall thickness, pipe diameter, width of the clamp, and size of the opening. PVC clamps are quick to use and inexpensive to make—perfect for epoxy.

Rubber strips—Rubber strips cut from discarded tire inner tubes (which you can often get for free from tire or bicycle shops) can be used to tightly wrap a bundle of glued strips. You cut the strip like peeling an orange. Wider strips offer more pressure, narrower strips less. While rubber strips are fast to apply, they offer only low clamping pressure.

More often than not I apply the rubber strips to the bundle after I apply the glue, to close the glue joint. Yes, it's extremely messy. Minimum "open time" is critical to the success of a joint. Once I wrap the bundle, I bend it around the bending form and clamp right over the rubber.

Removing the rubber can be quite a chore. The rubber sticks to the glue better than you want it to. If you use epoxy or plastic resin, the correct glues for this process, beware of sharp points of hardened glue. This is always a time when I am sure to wear glues.

Strapping tape—Strapping tape is surprisingly effective for clamping. It may be used to tightly wrap the bundle of laminates the same way as rubber strips. Even if you use it only here and there, it's good for holding the bundle together so the laminates stay in full contact with each other during the bending and clamping process.

This bundle of veneers has been stacked just the way the pieces were cut from the log. Now when a bundle of them is glued back together for lamination-bending, the grain will match.

VENEERS VS. MILLED LAMINATES

The first and usually best choice when lamination-bending is to use sliced veneers for laminates. They are just the right thickness for almost all bending. They offer the best overall quality of bend because not only are they flat and thin, but the grain match is perfect as well. There is almost no waste because the veneers are sliced from a log into a flitch—a bundle of veneers cut from a single log, stacked on top of one another to reform the shape of the log.

If we bend something like a chair hoop that is 1 inch in diameter, a bundle of veneer approximately 1-1/8 inch square will be needed for bending. This allows for some trimming, sanding, and shaping. You determine how thick and how wide the piece being bent must be, then cut a stack of veneer accordingly. The process ensures a laminated bundle where grain and color are so perfectly matched that the assembly looks like a solid piece of wood. The common veneer thickness, about 1/16 inch, will be flexible enough to accommodate all but the tightest radius.

Ripping a stack of veneers from the bundle makes short work of precision bending. The veneers are thin enough to bend to almost any radius. More importantly, once the glue-up is sanded smooth, the grain match is perfect.

Milling your own laminates

Milling laminates yourself and keeping track of the pieces so that the grain matches up will achieve the same results as using commercially sliced veneers. In addition, you'll have the advantage of being able to mill the laminates to any desired thickness. The downside is that milling your own veneers can be dangerous and wasteful.

Cutting laminates on a table saw

Using a table saw to mill laminates is fast and accurate, but dangerous. The commonly used process has the rip fence in the table saw too close to the blade, preventing the use of most of the more common safety devices. Also, if thick lumber is being cut, the blade will be set quite high, a hazardous undertaking in itself.

By moving the rip fence each time, bending layers can be cut easily and safely. Mark the top of the board with a pyramid mark to keep track of the bundle so that the grain matches. A good table-saw cut is perfect for gluing.

To make table-saw operation safer, use an after-market splitter. It's foolish to run a table saw without one. To rip laminates, I suggest you first cut both edges of the board straight and parallel. Then move the rip fence over toward the blade for each cut so that the offcut is the piece you use for the laminate. When the workpiece gets so small that your pushstick no longer fits between the fence and the guard, start on the next wide piece.

Here's a good way to keep track of the cut pieces so you can assemble them in the right order. Just draw a triangle across the board's width and another on its end before making any cuts. It will then be obvious which pieces go together and in what sequence. Handle the pieces carefully to avoid damaging the edges.

Cutting veneers on the table saw is a snap with a power feeder. The feeder keeps the cut going smoothly and evenly, resulting in a surface that's ready to glue.

Consider adding a power feeder to the table saw when milling your own laminates. Not only will you boost speed and accuracy, power-feeding is much safer than hand-feeding. The feeder, which bolts to the saw table, holds the wood down and pushes it through the cut at the same time. Since the piece is automatically fed past the blade, the operator's hands stay well clear at all times. In addition, the feeder covers the blade area, helping keep chips from flying in the operator's face. But the biggest advantage of a power feeder is the quality of cut. The constant feed rate produces a surface that needs no further milling prior to gluing.

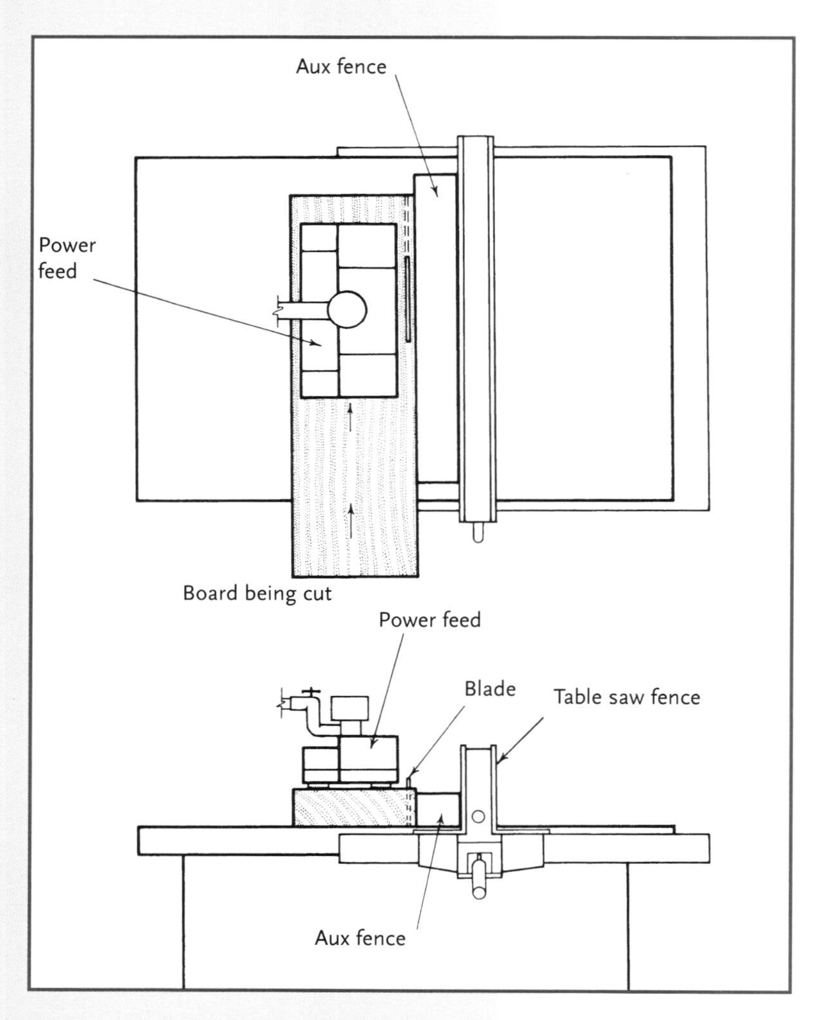

When using a power feeder, the auxiliary fence should be lower than the piece being cut by about 1/4 inch.

In order for a power feeder to work well, the table saw must be in tip-top condition, have a sharp blade, and at least 3hp. Using a power feeder requires a secure rip fence as well, since the feeder puts a great deal of pressure on the fence.

For the feeder to work, the fence cannot interfere with the pressure of the feeder. Attach an auxiliary fence to position the fence out of the way. Spray a lubricant like Top Coat on the table top and auxiliary fence. Even though the feeder will tend to hold the workpiece against the fence, I always use a feather board just in case.

Cutting laminates with a band saw

Initially, it seems more logical to cut laminates with a band saw than a table saw, since the kerf is usually narrower and it seems like there would be less waste. In fact, it works out to be a wash between the band saw and the table saw.

If you joint one edge of a board then band saw it, the rough face must then be run through the thickness planer for smoothing. In addition, the rough band-saw cut must also be jointed smooth. Combine these two operations and the table saw looks better all the time.

One exception is when cutting very wide laminates. The tallest material you can cut on a 10-inch table saw is around 2-1/2 inches. Any thicker and you must use a band saw. If you plan to rip a 6-inch thick board into thin veneers, the band saw is a good—and possibly the only—alternative. The procedure would be to rip one laminate, plane the cut smooth again, then rip the next cut. All of the laminates would need to be surface-planed to smooth the band-sawn surface.

Preparing the surface of the laminates for gluing

Proper surface preparation is critical to a good bond. If a surface is dirty or glazed, no glue will bond well. Remember that the bonding of the laminates is the only thing that keeps them in the bent state.

In my shop, we routinely cut laminates on the table saw and glue them straight from the saw. The cut from a well-tuned table saw with a good blade cannot be improved upon. In my opinion, if you're not getting a glue-quality cut from your table saw, perhaps it's time for a tuneup.

If you are cutting on the band saw and resurfacing the wood on a jointer or surface planer, make sure the knives are sharp and that they're not glazing the wood surface. Test this by dropping water on the wood surface after it's been run through the thickness planer or jointer. The water should promptly soak in, not stand on the surface.

If planing or jointing, you must also allow for snipe. Snipe is where the jointer or surface planer cuts into the first few inches of a board more deeply than the rest of the cut. There is no real solution for this problem, as all surface planers do this to some degree. About all you can do is allow enough extra length so that the snipe can be trimmed off.

GLUES FOR LAMINATION-BENDING

Good choices are plastic resin, resorcinol, or one of the boatbuilding epoxies. These glues become rock hard when cured, and hold the laminations together without slippage or "creep." Glues like these are designed to cure much harder than common white or yellow woodworking glue. Regardless of the glue, read the directions on the container. Follow them. Clamping pressure, temperature, open time, pot life, and shelf life all vary according to the type of glue and the manufacturer.

Plastic Resin

Plastic resin glue is usually my first choice. It normally comes as brown powder in a plastic tub. It's inexpensive and, when mixed with water, has a long pot life—you can go have lunch and when you come back, the glue will still be ready to go. Plastic resin glue doesn't fill gaps very well, but seems hard as glass when cured. A negative is that the dark brown color of the cured glue can detract from laminations made from light-colored woods such as maple.

Another form of plastic resin glue is modified urea formaldehyde, which goes under the brand name of Unibond 800. It uses a similar powder, but instead of mixing with water, you mix with the liquid resin supplied. Much like resorcinol, the glue is waterproof and extremely hard. Unibond has the advantage of a light brown color and is formulated to help fill any gaps as well.

When using this glue—both in its powder and liquid forms—carefully follow the safety precautions described on the container. Wear a dust mask when

Plastic resin glue is the workhorse of lamination bending. It mixes with water to form a strong bond suitable for pieces under pressure.

Modified urea formaldehyde resin glue is a two-part adhesive that's perfect for gluing panels that will be under stress during lamination-bending.

mixing and sanding the cured glue to prevent inhaling glue particles. Keep in mind that plastic-resin glue is hard enough to damage edge tools like planer blades and jointer knives.

A beater from an old mixer in a cordless drill works well to stir the powder and water together. Start with a healthy slug of water to minimize lumps.

Mixing powdered glue

Since you're working with toxic stuff, wear gloves, long sleeves, a mask, and safety glasses. Try to stir as little as possible until the powder is completely absorbed in the water. Otherwise the powder will fill the shop with a toxic cloud.

It's difficult to mix urea. I use a beater from an old mixer chucked in a cordless drill. I start with as much powder as I figure I'll need, add a fairly large amount of water in one shot (roughly half the measure of the powder), then mix the glue to a paste-like consistency, making sure all the powder is absorbed. I gradually thin the mixture with small amounts of water, mixing thoroughly until the consistency is just right—similar to heavy cream. Remember that you have the luxury of time when mixing plastic-resin glue, since it has a pot life of several hours.

The glue will almost always be too thick or too thin. Add either small amounts of water or powder until it's about the consistency of cream or syrup.

The mixture will always be at least slightly lumpy, so you will have to strain the glue. I recommend an old nylon stocking: Cut a section about a foot long, tie a knot in one end, stretch the other end over a coffee can, and pour the glue through. It's easy to force the remaining glue through the stocking by squeezing it, the way you would milk a cow. You did put gloves on, didn't you?

Epoxy

Boatbuilding epoxies are specially formulated to glue bent laminations. They need little clamp pressure, fill gaps well, and cure to a clear, light, amber color similar to that of varnish. The color of boatbuilding epoxy makes it a logical choice for applications using light-colored wood or those that will be exposed to moisture. But it's extremely temperature-sensitive and is one of the more expensive glues to purchase.

A piece of nylon stocking stretched over a coffee can makes a good strainer. Wearing gloves, force the glue through the nylon. Clean out the mixing pot and pour the lump-free glue back in.

Mixing and measuring boatbuilding epoxy couldn't be easier. For example, the West System brand of epoxy comes in two separate cans with pumps—one pump of resin and one pump of hardener give you the

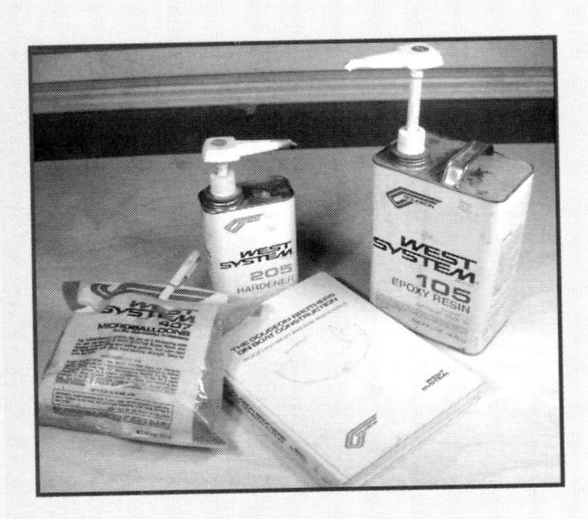

Two-part epoxy glue was formulated especially for boatbuilding. It works well for nearly all woodworking applications that are under stress.

exact 5:1 ratio you need. Simply stir the resin and hardener together and that's all there is to it. This glue works exactly as the literature describes, but you must follow the instructions, including safety precautions, exactly.

One of the reasons I like West System brand epoxy so well is that a support structure within the company is there to help in case you encounter a problem. Available from the company are a full-sized text book on using the epoxy, pamphlets on specific uses of the product, a newsletter giving you the opportunity to contact others working on similar projects, a website, a video, and a telephone hot line.

A word of caution if you've never used this material before. A chemical reaction begins the moment the two parts of the epoxy meet. This reaction generates heat. The heat builds up quickly if left in the mixing pot and can get hot enough to melt through a Styrofoam cup in just a few minutes.

Once mixed, the glue sets up in just a few minutes, ready or not. The good news is that the literature tells you exactly how much time you have based on the type of hardener and the temperature in the gluing area. (Different hardeners are sold to work with a variety of weather conditions.) The literature is accurate to within a minute or two for a mixture left in the mixing pot. If you don't pour out the glue or use it quickly, the mixture will get hot enough to start smoking. If this happens, carefully pour the mixture into a metal pan and let the heat dissipate. This might save the batch of epoxy, but if you see any sign of thickening, mix a new batch rather than risk getting your bundle of strips half-glued only to discover that your glue is setting up.

The trick to extending the pot life of epoxy is to keep it cool. If you pour it into a cookie sheet or paint rolling tray after it's mixed rather than leave it in its mixing pot you can easily double the normal time after mixing. Spread it out so only a thin film covers the sheet.

WORKING WITH GLUE

What's the best way to spread glue? The short answer is "quickly." This is a good time to get used to the notion that it's easiest and best to glue all of the laminations at once in an oozing bundle of strips. It's the only way you'll be able to get consistently good results. To pull this off successfully, it's obvious you must work speedily.

Working within the open time of the glue

"Open time" is the period of time the laminates stay open to the air after you apply the glue. From the time you roll on the glue until the two pieces come into contact, you must work quickly. You want the absolute minimum open time. In some warm environments, this means only a few seconds. One of the reasons I like to spread on lots of glue is that by doing so, I slightly extend the length of time the joint can stay open to the air. The thicker coating of glue doesn't seem to skin over as quickly as a thin coat, so it gives me a bit more open time.

The entire process of spreading the glue and bringing the pieces into contact with each other must be done within the open time. Spread the glue on all the pieces, stack them together, and hold them tight with either packing tape or strips of rubber. Then take this oozing bundle over to the form. You still have to work quickly because now the clock is ticking on the glue's working time. The working time for glue is the amount of time you have to get the bundle on the form and completely clamped. Again, it's important to read the literature from the manufacturer, but fast is good. The more quickly you can spread the glue, close the pieces together, and clamp the whole business, the better.

If you have a joint get away from you by working too slowly and the glue skins over, here's what to consider doing about it. Before I go much further, I must say that when this happens in my shop, I start over with new material. But I realize this goes against the grain sometimes. So if you're determined to use the glued material, here's what to do.

First, scrape off all the glue you can get off. Then re-mill all of the pieces once the residual glue has cured. This means removing a layer of material from both surfaces. Run the layers through the surface planer, making enough passes to completely remove all the old glue. Remember that the glue is quite hard and may damage the knives. And since the layers are now thinner, you may need to add an additional laminate or two to achieve the total thickness you want.

THE PROBLEM WITH WHITE AND YELLOW GLUE

Common PVA woodworking glues remain slightly flexible even after fully cured. This makes them unsuitable for lamination-bending, since a lamination bend is under constant stress. Over time, given this stress, soft glues like yellow and white glues may allow the bend to straighten out, or creep. Over time, a bent lamination under stress will cause the bent part to slide back toward its natural straight form. ■

THE SHELF LIFE OF GLUE

Here's something to remember—most of the suitable glues for lamination-bending have very limited shelf lives once opened. A container that has been around for too long may be too old to use. Modified urea formaldehyde resin glue, for example, has a shelf life of three months if stored at high temperature and will last but one year in normal temperatures. Urea powder should smell strong—like pine disinfectant. If the smell doesn't bother you when you open the container, it may be too old.

Boatbuilding epoxy, on the other hand, has an almost unlimited shelf life. Although the hardener tends to darken with time, bonding properties don't seem to be affected. ■

Glue must be applied quickly. A roller saves time even on a small project.

Apply glue to both sides to ensure full coverage on both sides of the laminates. Don't depend on one piece to transfer glue to the other. Do that job yourself with a roller.

This is the way liquid glue should transfer to your finger. I always apply glue to both surfaces and make sure the pieces don't stay open long enough for the glue to skin over. You must get used to working quickly. A dry run is the best way to speed the actual gluing operation.

This glue joint is not going to work. I've waited too long and the glue has skinned over. The glue must be liquid when the pieces come together. If you don't have glue sticking to your finger when you touch the open joint, you must start over again from scratch, re-milling the pieces.

Glue can skin over in a matter of minutes, especially in a warm environment. The skinning process is greatly slowed once the laminates are in contact, since the joint is then closed to the outside air. Beware, however, of two layers with a gap in between them—the gap allows air to reach the glue, which can eventually cause delamination.

Plan to tape the oozing bundle together as quickly as possible with fiber packing tape to keep the bundle from opening before you get it clamped. Spreading the glue on the laminates with a narrow paint roller greatly speeds the gluing process. A heavier coat of glue is preferable to a lighter one. The lighter the coat of glue, the faster it skins over. This is not the time to scrimp on glue. Remember to coat both sides of each laminate with glue. Why? Take two pieces of scrap. Coat one surface. Place the pieces together. Now pull them apart. It's easy to see how spotty the glue transfer is. Don't take a chance. Coat both sides.

There is no substitute for a dry run to make the process go faster. Prior to gluing, practice laying out the laminates edge to edge and bundling them together a couple of times. Since it's extremely easy to get a laminate upside down, check that your marks are all on the top. You've gone to considerable trouble by this point to keep the grain aligned, so it's nice to get all the pieces right side up the way you intended. Now is also a good time to remember that the two outside laminates are coated on one side only. Make sure to mark the gluing surface.

There might be a reason to have one side of your bundle be the concave side and the other the convex: If the bundle must go on the jig in a certain way, mark which side goes where. As you pick up the oozing bundle, you'll want to know just how it should go on to the form without having to think about it.

Another sure way to improve your speed and technique is to clamp the bundle of laminates onto the form without glue. This dress rehearsal will test your clamping system and make sure everything you need is at hand. It will also serve to calm some of the tension normally associated with bending. You could well find that you are two clamps short of the number you need. The time to borrow them from the neighbor across the street is before you spread the glue, not after.

Bending and clamping

This rather messy process means that you have plenty of glue.

Let's assume that you're ready to go. The glue is mixed, the form is ready, you've dry-clamped the laminates in place. You're back from the home center with the clamps you didn't think you'd need. You've vacuumed the laminates (glue doesn't adhere well to dust.) The moment of truth has arrived. This is the time to turn on the answering machine, take a bathroom break, and put the dog in the yard. The moment you start to spread glue, you are committed to finishing the bend.

As I've stressed endlessly, when you start to work, fast is good. Once the glue hits the laminate, it starts to set. There is no way to overemphasize the importance of speed while spreading the glue, bundling the laminates, and getting the laminates onto the form. Once you're done with gluing, bundle the laminates using strapping tape around the bundle every couple of feet. It won't interfere with the bending and will serve to keep the laminates from separating during the process. It also makes the bundle easier to carry.

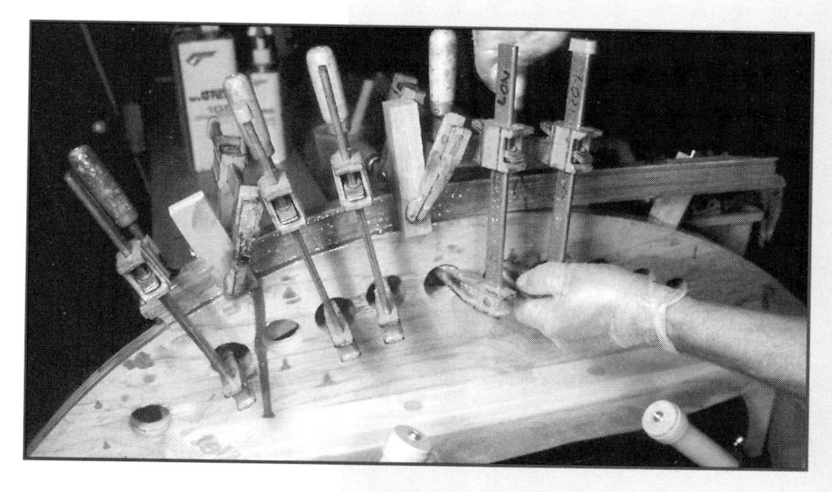

By starting at one end and working toward the other, the layers are squeezed together evenly and excess glue is squeezed out. The same result is possible by starting in the middle, then working out toward the ends. What you don't want to do is start at the ends and work toward the middle.

Your dry run should have told you how best to proceed with clamping. But expect the unexpected. The oozing glue always manages to change the way the actual glue-up plays out.

You can start clamping from one end and work toward the other end, or you can start from the middle and work outward toward both ends. What you don't want to do is start at both ends and work toward the middle. Invariably, what will happen is that a gap will open in the middle, separating the layers instead of closing them. The bend should close gradually, uniformly forcing the excess glue out from between the laminates as you go. The bending should happen in a way that allows the

IF YOUR SHOP IS COLD...

If your project is fairly small and it's not feasible to heat your shop to the required temperature, it's quite possible to still glue quite satisfactorily. You can make a tent using an old blanket or sleeping bag, place all of your gluing materials in it including clamps and glue, and put a lamp with a single light bulb in the tent overnight. The next morning, glue and clamp, leaving the glued-up assembly in the heated tent for an additional 24 hours. ■

laminates to slide freely on one another.

You should not need to grit your teeth to tighten every clamp. Light, firm pressure should allow the excess glue to flow out of the joints while forcing glue into the wood pores. Too much clamp pressure can cause flat spots and, in forcing out too much glue, starve the joint. Clean up excess glue from the workpiece, the benchtop, your shoes, and anything else while the glue is still liquid. I don't recommend a wet rag for cleaning up excess glue. The last thing you want to do is dilute the glue where it matters most, that is, at the most visible portion of the glue joint, right on the top. Scrape off the excess with a scraper, and leave it at that until it's time to unclamp.

Twenty-four hours of clamping time should be plenty for any glue. Make sure that your gluing environment remains within the temperature recommended by the glue manufacturer for the entire time. If the temperature drops, the glue may not cure properly.

Glue cleanup

Sanding or planing off excess glue can be hazardous. Many people are allergic to the chemicals in these glues, especially in the epoxies. When sanded in the shop without proper dust collection, cured glue produces a cloud of toxic powder. If the operator is unprotected, the powder will quickly collect on the skin and in the lungs, which will readily absorb the chemicals. Obviously, the best line of defense is to capture most of the dust before it gets into the air, and protect yourself against the dust that is bound to escape.

I don't recommend removing excess glue by running the piece over the jointer. The beads of glue form an irregular surface that may catch on the jointer throat or table edge, making smooth feeding of the piece difficult and hazardous if the piece lurches forward. Moreover, feeding a curved piece across the knives could well put your hands too near the knives. In addition, glues used for lamination cure so hard that they can nick the jointer knives.

Bear in mind that the glue beads can nick your hands, too. The beads can be sharp as razors—any sudden movement of the piece could

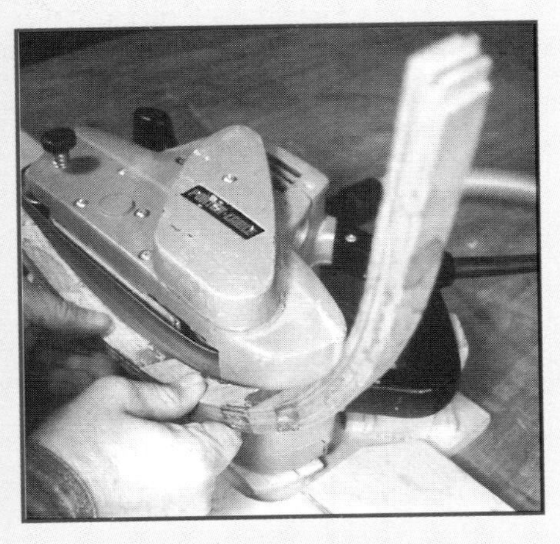

I routinely use a belt sander to smooth one side of the bend. Keep in mind the glue can be quite toxic, so use dust collection. The hose extending from the belt sander leads to a shop vac.

result in a serious cut, so wear gloves when unclamping.

I typically remove excess glue with a belt sander, electric planer, or hand plane. Hook the belt sander and planer to a vacuum to contain the dust. Long sleeves and a dust mask will add protection.

As an option, once one side is flat, you can carefully feed the piece through a small portable thickness planer to clean and flatten the other side. Use the kind of portable planer that has a flat metal plate below the cutter, not the type that has rollers—the planer must not be so large that the piece cannot be controlled throughout the cut. The planer will shave the upper side of the bend flat and keep the thickness uniform. Turn the piece over for one last pass to make sure it's flat without low spots. This technique will not work on a larger planer with bed rollers, only on a portable with a solid bed.

Keep in mind, also, that running these vertical glue lines through the portable planer may nick the knives. In my shop, I have two sets of knives for my little portable planer—one just for this purpose.

Sand the concave and convex faces of the piece to remove all traces of glue. Use an edge sander or a belt sander on its side to sand the convex face. The concave face is best done by hand. Be mindful of the fact that the glue is often harder than the wood. It's all too easy to sand through the glue and gouge the wood. Carefully vacuum the piece once it is sanded, prior to shaping it with a router.

If the piece itself is to be round, you can shape it using a roundover bit with a bearing in a router table. If you are working with a piece that is 1 inch in diameter, use a 1/2-inch roundover bit. Take gradual cuts, the first barely touching the part. A little at a time, raise the bit until it cuts the full radius, rounding the piece.

An inherent problem with these bits is that the bearing runs below the equator on the last cut, often distorting the radius by cutting too deeply. I solve this problem by not raising the bit all the way and hand-shaping the last of the radius.

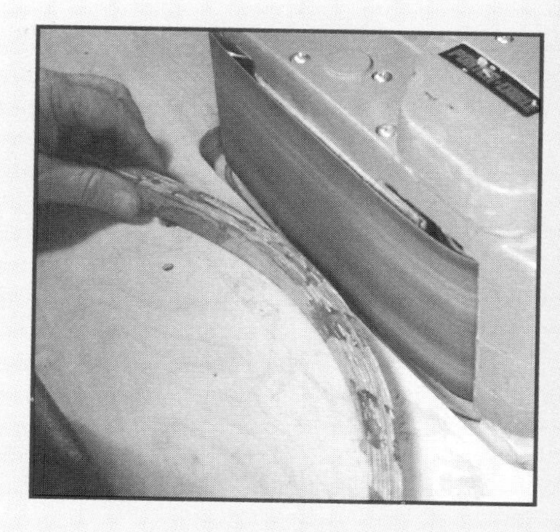

By holding the belt sander on edge, you can sand the outside and inside of the curved piece and the top and bottom as well. Once it's clean on all four sides, rout it into any shape you like.

Once one side has been smoothed with the belt sander, you can carefully feed the part through a portable planer. This makes the bent lamination smooth and the faces parallel.

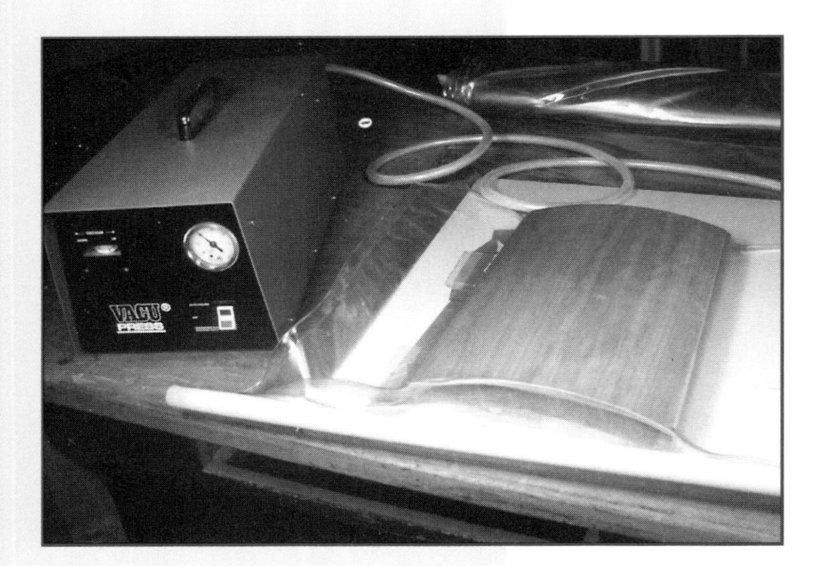

A vacuum bag holds the pieces in contact evenly until the glue sets—the ultimate way to clamp large curved panels.

CLAMPED CURVED PANELS WITH A VACUUM BAG

This fairly recent technology has revolutionized the kinds of curved parts and panels that small shops can produce. By facilitating even clamping over a large surface—as big as a piece of plywood or even larger—professional results with veneers and curved shapes can be quite straightforward. Here's how the process works for making a curved panel using a bendable plywood core and veneers on the outside surfaces.

The starting point of any project is the radius, length, and width of the panel. The layers that will make up the panel are glued up and clamped to a male form. The entire assembly—form, layers, and all—go inside a heavy plastic bag. A vacuum pump, which is the core of the system, sucks the air out of the bag. Atmospheric pressure supplies the clamping pressure evenly over the entire panel. The pump turns on and off automatically to maintain the vacuum until the glue sets up. Then the bag is opened, out comes the curved panel, and glue up is complete.

The layers

The number and thicknesses of layers depend entirely on the radius and the desired thickness of the panel. If the outside is going to be veneered, it's a good idea to glue up the core first, then, after the glue sets, glue the outside veneers in place as a separate step. Everything can be done at once, of course, but separating the steps makes the process a bit easier for the uninitiated.

As with any lamination bend, the more layers you have, the better. The outer veneers will bend more evenly if you first glue a perpendicular layer to the outer layer as a backer. This stabilizes the face veneer, flattens it, and allows pre-sanding if desired. It will still bend just fine.

If weight is a factor with a panel (as in making a cabinet for a boat or airplane, for example), a resin-impregnated honeycomb core material is available. Veneers glued to the two faces of the material make a panel that is curved or flat, depending on your requirements, very strong, and very light. Around the edges, solid-wood inserts the same thickness as the core material can be glued in place, bonding the veneers to the inserts, providing solid backing for hinges, and banding the edges all in one step.

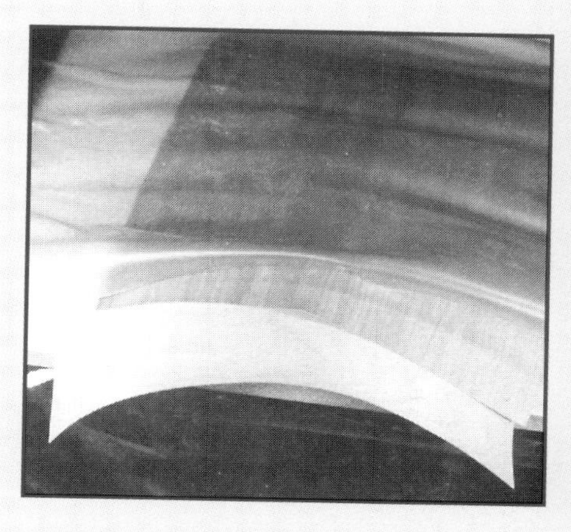

Gluing a perpendicular layer of veneer under the outer layer stabilizes the veneer and allows for a bit of initial sanding prior to gluing up the part.

The form

Since there is such a great deal of pressure with this system, forms must be strongly built. The typical form is made from ribs, like an airplane wing, skinned with plywood.

Bending plywood—Bending plywood is made of flexible poplar in either 1/8-inch, 1/4-inch, or 3/8-inch thick sheets. It bends so easily because the plies are thick in one direction, thin in the other. Depending on your project, you can order the plywood so that it bends easily along either the long (8-foot) or short (4-foot) dimension. Bending plywood does indeed bend very easily, but there are also other alternatives: See the Source of Supply Appendix.

The process of bending plywood is virtually the same as bending solid wood. You use laminates thin enough to bend easily around the radius. With a solid panel, say a cabinet door panel, you have many choices for core material—bending plywood is only one. The honeycomb-material mentioned above bends quite readily. Thin plywood (about 1/8 inch thick) will bend easily. The idea is to make sure the plywood core material is as flexible as it needs to be so that it can bend over the radius you are working with.

Vac-bag forms must use ribs that are closer together than with other clamping methods. The amount of pressure will put scallop-like depressions in the part if the form is not built really strongly.

The ribs—The ribs in a form used for vacuum-bending must be spaced closely together—not more than 6 inches apart. The form itself must have at least one and sometimes two layers of 3/8-inch thick bending plywood for the skin. If the form is underbuilt, when the vacuum is applied, it will sag between the ribs like a fabric-covered airplane wing. If a smooth panel is your objective, you cannot scrimp on form materials.

You can either cut the ribs carefully with a band saw (see radius cutting section, pg. 61), or you can carefully make a pattern and cut it to shape with a router. The bottom sheet is 3/4-inch solid plywood. The ribs are fastened to it first. Then the curved top skin is applied.

Gluing up the panel

Like any glued-up assembly under stress, panels glued up of several layers require adhesive that is rigid, not flexible. Epoxy and plastic-resin urea-based glues are suitable.

Before spreading any glue, take time to make a dry run. As part of the dry run, include getting the form and glued-up panel into the bag as well as clamping.

For an even coating, apply the glue to both surfaces of the joint with a roller. With porous materials like bending popular, which absorb glue, you need to apply a good heavy coat to ensure that enough will remain in the joint. Work quickly—as you must any time you spread glue in a lamination-bending process.

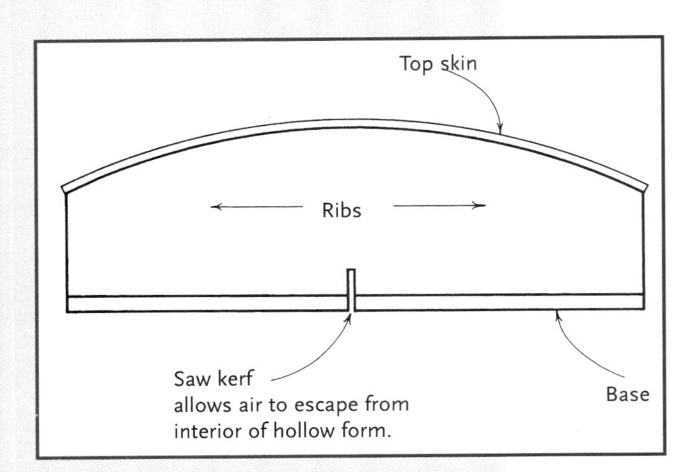

To allow air to escape from the interior of the panel, make a saw kerf along the length of the form, cutting through the bottom panel slightly into the ribs.

Getting the assembly into the bag

This can be quite tricky if you are working in a confined space. A manufacturer of vacuum systems recommends placing a grooved melamine panel in the bottom of the bag to facilitate the evacuation of the air. The panel is easy to make from an old sheet of melamine. Simply cut 1/4-inch deep saw kerfs in a grid pattern over the entire sheet. In addition, the panel makes it somewhat easier to slide the dripping assembly into the bag.

Once the assembly is in place, make sure it is centered on the form, not twisted to one side. As the air is sucked out of the bag and clamping begins to take place, make sure the bag contacts the panel evenly so that the lower edges are pulled into the form. If the glued-up panel gets rotated, the corners will hang off of the form. The vacuum pressure will distort the overhanging edges.

TROUBLESHOOTING A FAILED LAMINATION BEND

A failed glue joint in a bent lamination could be embarrassing. If you happen to be a professional woodworker, it can be much more than embarrassing. It can be a disaster.

It's always easy and tempting to blame the glue, but alas, the glue is rarely the real cause of trouble. Operator error is much more likely. If you do have a failure, try to pinpoint why. Look to the following for clues in order of likelihood.

Too much assembly time, especially open time in warm weather

If too much time goes by between when the glue is spread and the time the pieces are assembled, the glue may skin over like a blister and prevent a good bond.

Wrong type of glue

White and yellow woodworking glues are great for a lot of things, but lamination-bending is not one of them. The very flexibility that makes these glues so tough in other woodworking projects may lead to eventual failure under the stress of a bend. It all depends on the amount of pressure. If you use 1/16-inch thick laminates to bend a 1-inch thick bundle over a 48-inch radius, it is unlikely you'll have trouble even if you use yellow glue. But if you force laminates that are thicker than will easily bend around the radius you have, you're inviting trouble unless you use epoxy or plastic-resin glue.

Poor surface preparation

Poor surface preparation is a leading cause of glue failure. Dust on the gluing surface; dull cutting tools, which can char or glaze the surface (both of which impede the ability of the wood to absorb glue); and surfaces that don't fit together very well all contribute to sabotage the job the glue is trying to do.

I either vacuum or use compressed air to clean dust off wood surfaces. Imagine trying to glue two boards together with a layer of sand between them. The glue would stick to the wood just fine and to a few grains of sand as well, but it would not be able to bond the two boards.

Once a board is cut, the surface immediately starts to oxidize. The yellowish tinge most woods acquire after a few months sitting around in the shop is just as effective a sealer as dull jointer knives. So always use freshly milled gluing surfaces.

Temperature is too low for the glue to cure

If the temperature in the shop is too low, the glue may not set. Some glues are forgiving in this regard, but most are not. A careful reading of the literature will give you the temperature range. If proper temperature is not maintained, failure is not the fault of the glue.

Poorly fitting joints

Laminates must have flat gluing surfaces. The ripples often left behind by poorly tuned planers and jointers may cause pieces to fit poorly. The better the fit, the thinner the glue line and the more effective the glue joint. The exception is epoxy—this glue doesn't care how far apart the two surfaces are.

Old glue

Glue does go bad. It does not last forever sitting on the shelf in your shop. Even if it is still in powder form, it may have gone bad. It's good practice to write the purchase date on your glue containers and find out how long the glue can be stored without going bad. if you've had glue longer than a year, it's suspect.

Conclusion

It's safe to say that there are a lot of ways to get into trouble with glued lamination-bending, but every day people successfully bend projects by the hundreds, from furniture parts and handrails for circular staircases to wooden boat components.

Most common pitfalls, like using the right glue or getting the right thickness of the laminate, are easy to avoid. Other pitfalls are less obvious, like allowing for enough over-bend to compensate for springback or staying within the proper temperature range for the glue. By following a few steps, each of which is fairly simple, good results can be expected right away.

Of course there is no substitute for practice, but modern glues, machinery, and techniques make the process pretty straightforward.

PART FOUR

Steam-Bending

The wood in the downward-bending section of this stair's handrailing was steam-bent. Different curving techniques were used to form other components, as discussed in Part 5.

Steam-Bending

SEVERAL YEARS AGO I was commissioned to design and build a very different-looking staircase. My clients wanted something contemporary with simple lines running parallel to the incline of the stair. When I began to draw the stair, the design seemed to come out of nowhere. My staircase looked good on paper. My clients loved it. But how in the world would I build it? I assured them that, of course, I *could* build it. Just as they were about to sign on the dotted line, they asked how I was planning to make the quite prominent curved areas on the handrails. I replied, overconfidently, that of course I would steam-bend them. Little did I know that what I had just said would heat my living room that winter with an abundance of half-bent, split, broken, and generally ruined premium white oak that I might just as well have purchased by the cord.

Had I ever done any serious steam-bending before? No. I had experimented with bending a few times but without success. I kept thinking that if Norm Abrams did it on his television show, how hard could it be? I was about to find out.

The good news was that my clients had, by chance, selected an ideal bending wood, white oak. The bad news was that I wouldn't know for several years that the carefully kiln-dried lumber from the local hardwood center would offer considerably more challenge than bending the still-green and carefully rived oak used in the TV show.

And while the fact that my clients really liked my design concept was good news, the bad news was that my design called for bent pieces up to 5 inches wide by 1 inch thick. These had to be bent over a 7-inch inside radius to about a 45-degree angle. It would take me weeks of trial and error to discover that this was a severe bend indeed.

I was in for the steepest learning curve of my woodworking career. Like many woodworkers, I had played around with some steam-bending before, but nothing like this. I tried everything I could think of but nothing worked. I would hear this odd crackling sound every time I attempted to steam-bend a piece. This sound, as I would learn, signifies the limit of the bend. Whether or not the bend was as tight as it needed to be, this was as far as the wood was going to bend. The crackling sound was made by wood fibers starting to part company. It was as much warning as I would ever get that the piece was about to break. And break the pieces did—one after another. The crackling sound would soon be followed by a string of explicatives and, shortly thereafter, the crosscut saw would be heard making sure yet another piece of lovely white oak was short enough to fit in my fireplace.

In this section, I'll paint a picture of what steam-bending can and cannot do. Remember that we've already discussed my preference for milling curved pieces from solid stock. My second preference is lamination-bending. A distant third choice is steam-bending. While steam-bending seems like it's the answer to every curved project, in fact the applications for steam-bending are limited. You will soon realize that this is a specialized method for making certain specialized pieces, not a general catch-all technique to use any time you have a curve to make.

Steam-bending either seems impossible or easy, depending on how it's going. Rarely is there a bend that almost works or works pretty well. The learning curve is such that the bending usually either works perfectly or does not work at all. The goal in steam-bending is to control the variables one at a time. Your results will improve when you become familiar with the principles of bending. As soon as you experiment a bit, your expectations will become more realistic as well.

ALTERNATIVES TO STEAM-BENDING

Some of the time you can get wood to bend without steaming. Soaking the board, putting additives in the water, or simply finding a board that will bend enough without heating are all routes you can take to get a piece of wood from point A to point B. But don't get your hopes up. It's rarely the case that these methods work.

Soaking the wood—Soaking dry wood in water swells and softens its fibers and somewhat increases its plasticity, allowing the fibers to crush together more readily. This decreases the stretching pressure on the convex side of the piece during bending. While soaking is sometimes enough to do the job, usually the piece will break before the bend is finished. There are two downsides to soaking wood. One is discoloration, produced when the minerals in the water (or the container) react with the wood. The other is the along-grain checks that can develop as the wet wood dries.

To get a feeling for working soaked wood, leave a couple of bending pieces overnight in a plastic pipe that is filled with water and capped on both ends. Pull out a piece and try to bend it the next day, comparing its flexibility (and any discoloration) to a dry piece. You likely won't see much difference.

Additives—Fabric softener and additives such as ammonia and meat tenderizer are sometimes added to the soaking water in an effort to make the wood more bendable. The problem with additives is that they become part of the larger woodworking project. An additive stays with the wood, going right along with it through every milling, gluing, and finishing process. As if there were not already a sufficient number of ways to have a glue joint or wood finish fail, add to the mix a foreign chemical like fabric softener and you could be asking for trouble. Imagine a phone call to a glue manufacturer inquiring about the failure of a glue joint with wood that was previously soaked in meat tenderizer.

If you want to experiment with additives, take a sample piece of wood through all the steps in the process, from soaking and bending to gluing and finishing. If your results are satisfactory, you can proceed accordingly. ■

BENDING THEORY

In order to bend, the surfaces of the piece being curved must change lengths. How can this be? While at first this rather confounding notion seems impossible, if we look at something already bent, the outside, or convex, surface will be longer than the inside surface by more than you might think. To bend a straight piece of wood, the outside surface must stretch, the inside surface must compress, or both actions must occur. There is no other way for it to bend.

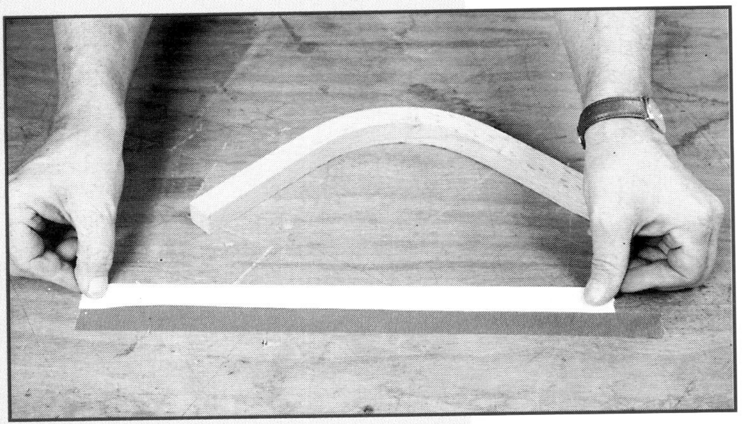

These strips of tape illustrate how the convex surface of a bend stretches and the concave surface compresses, resulting in different lengths.

Need proof? Find something bent—a chair part, a faucet, a table leg, or even a garden hose. Carefully run a piece of masking tape along the inside, or concave, part of the bend. Starting and stopping at the same place, run another piece of tape around the outside, or convex, part of the bend. Now remove both pieces of tape and compare their lengths. I'll wager that if you hadn't applied the tape yourself, you would suspect that someone was playing tricks on you. For purposes of comparison, consider that a 1-inch thick oak board bent over a 12-inch radius into a U-shape will show a difference in length between the two surfaces of over 3 inches. Picture having to shrink or stretch a piece of wood that is 1 inch square by 42 inches long by this amount and you'll begin to appreciate the forces at work when bending.

This chair crest rail may be steam-bent fairly easily by simply making a female form and bending the piece unassisted by a bending strap. A large radius and relative thinness both add up to a gradual bend.

If we set out to lengthen a board like this by pulling it from its ends, we would eventually pull it in two. The wood would break from tension failure. Simply put, this means that when you put the wood under tension, the fibers of the wood are likely to tear apart because wood is not particularly strong under tension. In fact, tension failure is at the heart of most bending failures. For a bend to work, the wood fibers must be allowed to slide past one another to lengthen the surface, but the stretching must be limited or the wood fibers will separate, causing the convex surface of the piece to fracture. By comparison, wood is much stronger under compression. This means that if you were to pile up weights on one end of a board, the wood, as long as it stayed straight, could probably bear several thousand pounds before collapsing. This concept seems far-fetched until you realize that this is what is actually happening on the concave side of a bend as a piece of wood is curved.

Left to its own devices, wood will try to bend by stretching only—a sure recipe for failure. Therefore, both tension and compression must take place for steam-bending to work. When the lignin in the wood is hot and soft, compression can occur because individual wood fibers are able to crush or slide into one another on the concave face. This allows the wood fibers on the convex face to stretch without being pulled apart by tension. Once you begin to grasp the forces at work, the steam-bending process becomes much easier to understand.

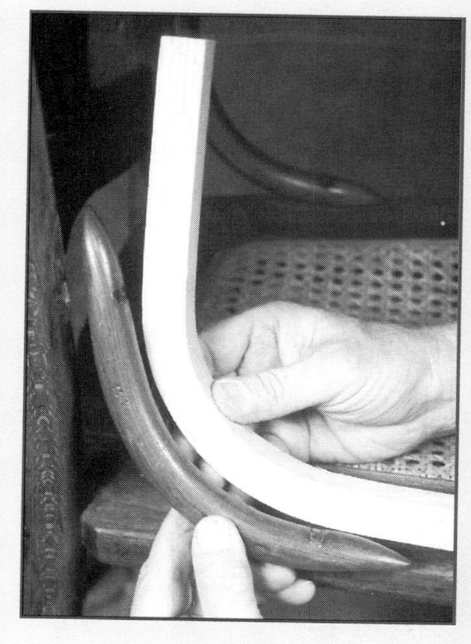

This chair bracket is a 1-inch thick 90-degree arc. The combination of its degree of arc and thickness make it much more difficult to bend than the crest rail shown on the previous page.

Judging the severity of the bend

As you can see in the photos on this and the opposite page, the difficulty of a given bend increases as the radius gets smaller, the degree of arc increases, and the thickness and width of the stock increase.

Compare a 12-inch inside radius with a 6-inch inside radius: Given the same 180-degree bend with the same 1-inch wood thickness, the bend will be concentrated in a much smaller area. The difference in length of the inner and outer faces of the two bends is about the same, but the length over which compression must take place is cut in half with the smaller radius, making the smaller-radius bend much more severe.

Obviously, a half-circle will take more force to bend than a quarter-circle. This difference in the degree of arc becomes apparent when you measure the difference between the inner and outer faces of the bend. A 180-degree bend requires that the wood compress twice as much as the 90-degree bend. And as the thickness of the wood increases, the difference between the inside and outside diameters increases, which requires additional compression.

In summary, as the wood gets thicker, the radius gets smaller, and the degree of arc increases, the bend becomes more difficult to control. As the bend becomes more difficult, controlling variables by using a compression strap (pg. 144) becomes more and more essential.

The hoop back on this chair qualifies as a severe bend. The wood is about 1 inch in cross-section, the radius is approximately 6 inches, and the arc is 180 degrees. The good news is that with a compression strap the bend becomes fairly straightforward.

ADVANTAGES OF STEAM-BENDING

There is usually no viable alternative to steam-bending solid stock in repair and restoration projects. If pieces were originally bent to shape, and you want to match the original construction, steam-bending is the only way to replace a broken part.

Unlike lamination-bending, steam-bending leaves no glue lines in the finished piece.

Steam-bending moves along quickly once the job is set up.

A steam-bent piece is virtually as strong as the unbent piece, although there will always be a slight weakening of the concave (compressed) surface.

DISADVANTAGES OF STEAM-BENDING

At least at first, steam-bending requires a lot of initial setup time and expense, is difficult to control, and the outcome is uncertain. Bends may be unstable over time—changes in humidity or direct exposure to water might straighten out a bend even years after the piece is built unless the bend is mechanically restrained. In addition, the final shape of the bend may not be precise. Lastly, water damage from the steaming process can mar the surface of the wood.

There are so many variables with steam-bending it is a wonder that the process ever works at all. Details such as wood species, time of steaming, temperature in the steam box, moisture content of the material, and the radius of the bend all effect the final outcome of a bend.

GETTING A FEEL FOR BENDING

The best way to get started in bending is to experiment with some small pieces of oak. This will give you a feel for the process and let you experience some of the variables before tackling a larger, more difficult project. When working with steam, be extremely careful. Refer to the safety information on pg. 142.

You'll need a camp stove for a heat source, a clean pot for a boiler, some heavy screen called hardware cloth, a piece of 3-inch plastic pipe about 3 feet long, and some aluminum foil. For bending blanks, gather several oak pieces about 1/2 inch square by 2 feet to 3 feet long.

Set up the camp stove with a pot of water. Cover the pot with a plywood lid having a hole the diameter of the pipe or a bit smaller. Put a piece of hardware cloth over the plywood lid; bend the hardware cloth around the edges of the plywood to keep it in place. Figure out a way to brace the pipe so it stays vertical.

Bring the water to a boil so you have lots of steam escaping from your little smokestack. With your gloves on, place three or four of the blanks into the steam chamber (the pipe). Cap the end of the pipe with foil. There's no need to try to prevent escaping steam—it should be blowing out all over the place. Note the time and write it down.

While the pieces are steaming, find some curved forms to use for practice. You can use a trash can or a coffee can—anything that is solid enough to try out different radii. While you're waiting, flex an unsteamed blank to get a feel for the natural bendability of the wood.

After 15 minutes put on your gloves again, quickly pull out one of the pieces, and recap the steam box. Flex the wood. Note that despite having been in a hot steaming box, the surface will appear completely dry by the time you replace the foil on the pipe. This illustrates how fast cooling takes place.

Now gently try to bend the wood over something. See how quickly it stiffens? Even though the piece is cooling rapidly and you must hurry, at the same time you must learn to move slowly and deliberately, since if you try to bend too quickly, the piece will break every time, as you will soon discover. Try to visualize the compression taking place on the concave surface and the tension on the convex surface. Give the piece some time to absorb these stresses.

See how far the wood will bend before it breaks. Go ahead and bend it until the fibers stretch beyond their ability to hold together. Remember that the piece wants to bend by stretching on the convex surface and will compress on the concave surface only when forced to do so. Notice what the wood looks like when it breaks. No doubt the break

A camp stove with a pot of water on it will generate enough steam to get started bending. Using a piece of PVC pipe for a steam box, supported above the boiling water by a plywood lid, it's possible to get a good feel for the steaming process quickly and easily.

For practice, start with small oak pieces measuring 2 feet to 3 feet long.

STEAM-BENDING RULES OF THUMB

These rules of thumb are only guidelines. Each bending situation presents unique challenges and will require different jigs and techniques. But taking the "rules" into account will help solve many of the common problems encountered in steam-bending.

Big is good: Picture a teapot on a camp stove with a tube running steam to a box 1 foot square and 12 feet long. How hot is it likely to get in the box? Not very. And without heat, steam-bending is not going to work well. How you set up a steam-bending operation will depend in part on how many pieces you plan to make. If you are making 100 chairs, you will need one type of setup. If you are building a set of four, you can get by with something less elaborate. It is important to proportion the project to the setup. Larger, thicker boards bent over tighter curves will need more steam, bigger clamps, etc., than smaller boards. Just remember that if you have to choose, bigger is usually better where equipment is concerned. It's also better when it comes to bending blanks. I always cut blanks somewhat larger than net size to allow for sanding and milling. It is usually best to bend the piece first, then sand and/or mill it clean afterward, since steaming, clamping, and handling often dent or otherwise damage the surface of the piece.

Setup: Most woodworkers either underestimate the difficulty of steam-bending or overestimate the capacity of wood to bend when steamed. If you are making 100 chairs, everything from clamps to the materials you use for the bending form will need to be more durable than if you were making only four, although there will be a similar learning curve. Regardless, make your setup bigger and stronger than seems necessary.

It's easy to underestimate the forces at work with bending. Remember that in order to bend, the lengths of the inside and outside surfaces must change in relation to each other. It takes tremendous force to do this. Also, prepare to customize your setup for each project. While you may reuse some of the same equipment, the boiler, for example, no one setup works for every situation. The form will be different for each piece. The box you use for a small piece may not accommodate a larger one. The compression strap may also have to be resized. The good news is that over time you will acquire a familiarity with these items so that building a new steam box, for example, will go quickly and easily. You may occasionally find that a used jig will work for a current application, but don't count on it. I say this now because you'll probably find that your first setup will not be your last.

Steaming: The boiler must be large enough to keep the entire box hot. The box temperature must stay at 200 degrees minimum. You don't have to worry that the box will get too hot, since nature limits the temperature in the box to 212 degrees at sea level. It's an interesting quirk that the pressure and temperature in the steam box are closely related. No matter how long you direct steam into the box nor how much steam you blow into it, as long as the pressure inside the box doesn't build up, the temperature inside will never exceed 212 degrees at sea level. This is a good time to mention that built-up pressure will cause the box to explode, which is why it's necessary to drill several vent holes in the box.

Use an ordinary meat thermometer to measure the temperature in the steam box. When all the pieces are in place and the box is up above 200 degrees, write down your start time.

The steam box: The steam box should be as small as possible. The smaller the box, the less energy it will take to heat it to the minimum 200 degrees. The box should be big enough for the material you are bending but no bigger.

It should also be insulated, which will help conserve fuel and make the box easier to heat. Shelves allow items to be separated so that the steam will reach all of the surfaces—I use dowels for shelves. Drill a few vent holes in the box so steam can circulate and escape. This circulation ensures that the temperature stays constant, with no hot or cool spots. Use a meat thermometer to check temperature at a number of the vent holes to make sure the temperature is the same throughout the box.

Time in the steam box: The rule of thumb is that a piece should stay in the steam box 1 hour for each inch of thickness. A 2-inch piece must stay in the box at least twice as long as a 1-inch piece for the heat to penetrate to the core of the material. But remember, this is only a starting point. A 1-inch square board will have heat coming at it from all four sides. A 1-inch thick piece that's 3 inches wide will have heat coming in from all sides, but two of the edges are much farther from the core.

Consequently the 3-inch wide by 1-inch thick piece will take a bit longer than a 1-inch square piece, but after an hour of steaming both will be close to being ready to bend.

The form: The bending form must be built strongly. It should facilitate quick clamping as the piece is being bent. In addition, the bending form must allow for over-bending to compensate for springback. It is always better to over-bend than under-bend. If the piece doesn't relax enough to achieve the correct shape, you can easily heat it again to straighten the bend. But while a bend is easily straightened a bit, once it's bent it usually can't be bent farther. Try to bend a board just an extra degree or two after it comes off the form and it will break almost every time.

Speed: Speed is essential to a successful bend. In general, you must have the piece bent within a minute of leaving the hot steam box. Once the wood is removed from the steam, it will remain plastic for only a few seconds.

Therefore, every second counts from the time the piece comes out of the steam box until the time when the last clamp is in place. While the pieces are in the box, try to make sure that everything is in place for the bend. Mentally rehearse the process. Put clamps where they will be within easy reach. If you have helpers, make sure they know their jobs. While bending must be done before the wood cools, the process should not be slapdash but rather slow and graceful. Seem like a contradiction? It's not really. The idea is to make haste slowly.

And remember, the farther you must carry the piece between the steam box and the form, the more your chances of success will diminish.

Compression strap: Most bending failures result from wood fibers being stretched past the breaking point on the convex surface. A compression strap limits stretching, forcing the piece to compress on the concave face instead. (As discussed on pg. 134, wood can more readily withstand compression than tension.) On severe bends, a compression strap is essential.

Practice with oak: No matter what material you intend to bend, practice with oak until you can bend it successfully and consistently. If a setup isn't working with oak, it's unlikely that another type of wood will work at all, let alone more easily.

This splintered piece simply lost its structural integrity. It failed because it wasn't heated all the way through.

This piece failed because it got overly softened and perhaps had a bit too much end pressure. The fibers collapsed, resulting in a piece that looks folded like a piece of paper. This is a compression failure.

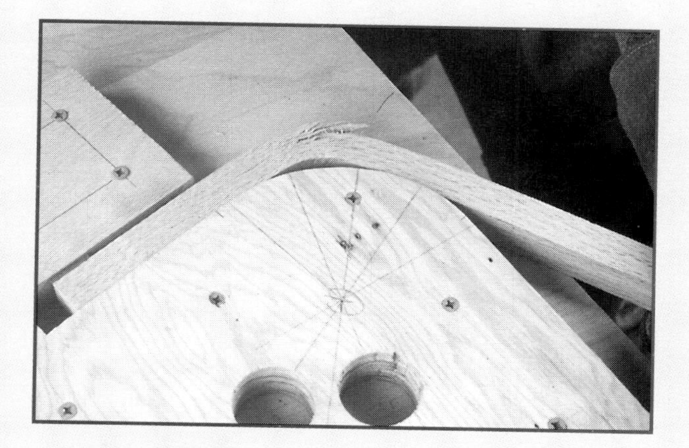

This form illustrates the futility of trying to bend a thick piece around a tight radius without a compression strap.

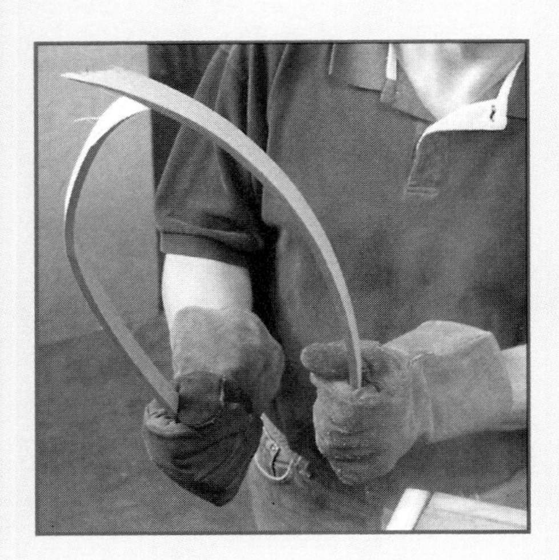

Wearing gloves, remove the foil lid and quickly remove the blank. See how far it bends before it breaks.

Grain runout is evident on these bending blanks. Notice the diagonal grain pattern on the blank on the left. It's best to select blanks that have grain running more parallel to the edges, like the blank on the right.

will occur at a point where there's grain runout.

While you've been playing with the first piece, the others have continued steaming. It's now probably about 20 minutes into the process. (That's why you recorded your starting time.) Gloves on, pull out the next piece. It will be slightly more pliable than the first piece. Try bending it, noticing again how quickly it stiffens. This is the very narrow window of time during which your chances of bending successfully are as good as they are going to get. Bend the piece to its breaking point, observing how and where it breaks.

You do get a slight warning before the wood fibers part company and the piece fails. It's a faint crackling sound—like cereal makes when milk is poured on it. Hear the sound and continue bending so you get the idea of how far you can go before the piece breaks.

As you bend, see if you can pick up different characteristics of the bending process. Does it work better if you bend quickly or slowly? Did the time spent in the box seem to matter? What improvements could you make in your boiler? In your bending form? In your material selection? In your technique?

Did any of your pieces have flaws? If so, you probably discovered that a flaw draws all the bending force to itself. If there is a small knot, for example, failure starts right there. Even cross-grain milling marks like those made with a band saw or surface planer sometimes initiate a break. The piece seems like it is just looking for an excuse to break somewhere. These slight flaws in the board substantially increase the chance of failure.

Grain runout

Trees are generally larger in diameter at the base than they are at the top. Most of the time, when lumber is sawn, the grain runs at an angle to the board, not parallel to it. By contrast, if wood is split from the tree, the split will naturally follow the grain direction. Since trees are normally tapered and planks are normally straight, it is somewhat rare for the grain to follow the plank unless it is cut that way on purpose. There is, of course, no substitute for splitting or riving a green log where grain runout is concerned, but we work with whatever material we have access to.

If you hold a piece to the light so that the light reflects off the surface, grain runout becomes quite obvious. If the runout is severe, beyond a slope of about 1:15, the fibers and cell structure will run at an angle to, not parallel to, the length of the board. This creates inherent weakness. If there is appreciable grain runout, breakage during bending will invariably start there. The sharp point of the grain as it runs off the edge of the piece will detach from the board as it is bent and total failure will quickly follow.

Sometimes it is possible to re-cut or even split a board along its grain to minimize runout, but usually this is unnecessary if you brought home straight-grain wood from the lumberyard to begin with. If you notice pronounced runout in a plank, cut it along the grain direction. Normal rip cuts will then yield blanks with relatively straight grain.

In addition to grain runout, any blemish, even a scratch in the wood, will give it the excuse to either break across the face in tension or collapse across the face in compression. Any sort of knot, flaw in the grain, or surface irregularity will tend to concentrate the stress of bending in that one spot. Failure will be almost inevitable.

HEATING THE WOOD

Wood softens just a bit when heated. It won't bend like a noodle, but it will become slightly more pliable. What happens is that the lignin, which is the "glue" that holds the wood cells together, softens as it heats and hardens as it cools. As the lignin softens, the fibers it holds together can compress slightly more easily than usual. The warmer the lignin, the more flexible the board, and the more plastic it will be.

For softening to take place in such a way as to make a board pliable, the heating must be as uniform as possible throughout the entire piece. If the outer layer of wood is hot but the interior is cold, the piece won't be flexible enough to bend successfully. The ideal temperature happens to coincide with the temperature of boiling water—212 degrees at sea level. The thicker the piece, the longer it takes for heat to penetrate to the center. The general rule of thumb is one hour of steaming for each

SAFETY

Most of us as children have run our fingers through a candle flame without being burned. If you were to try this with live steam, you would have blistered flesh hanging from your finger in an instant. If the steam were under even a slight amount of pressure, you would be lucky if your finger remained attached at all.

Live steam is totally invisible and incredibly dangerous. It will be at least 212 degrees; it could be much hotter if under pressure. On steam-powered ships, sailors are taught to walk along passageways holding out a broom handle to check for suspected leaks of live steam. If the broomstick is suddenly cut in half, they know they've found a leak. Remember, steam is visible only after it has cooled and condensed back into water vapor.

Boiling water in large quantities is always dangerous. This is a good time to send children and other curious onlookers to safer ground. Wear thick, leather welding gloves that extend to your elbows when loading and unloading the steam box and at all times when working around the box. Be mindful that as soon as you open the steam box a great deal of steam will escape. Stand back and let it stabilize before loading or unloading pieces to be bent.

(continued on next page)

inch of thickness. One of the reasons that green wood bends more readily is that the moisture in the wood conducts heat quickly and efficiently to the core of the piece. Remember, timing starts after the box is preheated, loaded, and back to bending temperature.

Although it's an oversimplification to say that it does not matter if the heat is wet or dry, the source of the heat is less important than the heat itself. The reason moist heat, or steam, works better than dry heat is because dry heat doesn't conduct the heat to the wood as well as moist heat. However, even moist heat will dry out a piece of wood if it is left in the box too long. The lignin will slowly cook right out of the wood.

Reheating the wood

While reheating a partially bent piece of wood may be tempting, it's probably a waste of time. If the piece cools before you get it fully bent, it's usually best to discard it rather than to try to reheat it, although you'll probably need to try this once just to see for yourself. Reheating means that the piece will be steamed for longer than it should be. The piece will dry out in the steam just like an overcooked Thanksgiving turkey. In addition, if the piece is partially bent, it will be difficult to fit it back into the steam box and harder still to get it into the compression strap. Having this problem says something about your technique. Either you are moving too slowly or your setup is too cumbersome. Perhaps you need to streamline your setup. Practice moving quickly when bending. Even though the actual bending is done slowly and steadily, getting the piece from the box to the bending form must be done speedily. Clamping must be done quickly as well.

Working with steam

Steam is wet heat at the ideal temperature—the perfect way to heat wood. Because it's wet, steam conducts heat effectively and transfers it efficiently, much more efficiently than dry heat. This is why it's important that the steam completely surrounds the wood in the box. Not only does this aid in the penetration of the heat, it also helps prevent cold spots in the board.

A steady temperature

At the temperature of boiling water, wood gets soft enough to bend yet still retains enough strength to withstand the bending forces without distorting or falling apart. There is no advantage in getting the wood any hotter—the fibers will begin to separate and break. It is possible to cook wood into mush with high enough temperatures in a pressure cooker. In steam-bending, the idea is to heat the wood just enough to bend without damaging the wood in the process.

You don't have to worry about getting the box too hot, since the water will get no hotter than 212 degrees at sea level. The real issues are preventing the box from becoming too cool and making sure that the wood is heated uniformly throughout the piece. I can't overstate the importance of a properly heated steam box.

What if the box is too cool?

We've seen that the box cannot get any hotter than 212 degrees. What if the setup is generating 150 or 180 degrees? Will it still work?

The short answer is no, but the actual answer is maybe. Lower temperatures may be adequate to make the bends for some jobs, but if your bend is more severe, it's likely that a cooler temperature will be insufficient. While you can't readily measure the core temperature of the wood, it's easy enough to measure the temperature in the box using a meat thermometer inserted into a vent hole.

If it measures below 200 degrees, there are several ways to raise the temperature. One way is to insulate the box and the tubes leading from the boiler. Rigid insulation wrapped around the box can raise the temperature by as much as 10 degrees.

But the best way to ensure a high temperatures is to supply more steam. If your current steam source is going full blast but not sufficiently heating the box, consider either a smaller box or a larger steam source. The size of the box must be in proportion to the heat source. The larger the box, the larger the heat source. Perhaps a second independent heat source could be used to supply steam to the box. Do whatever it takes to bring the box up to a minimum of 200 degrees and keep it there.

SAFETY (CONTINUED)

We will discuss various ways to generate steam, but all must be kept under control at all times. Every heat source represents a potential danger. Open flames are especially hazardous, of course, but even electric heating elements can overload a circuit and cause a fire. Don't use extension cords. Keep a fire extinguisher handy at all times. A helper whose job it is to monitor the steam generator is a good person to have around.

Make sure that the boiler never runs out of water. This could cause overheating. It's best to add water in small quantities frequently rather than in large quantities less often so that you can maintain the boiling temperature. Should the boiler run out of water, let it cool before adding more water. Never add water to a hot boiler that has run out of water. This could cause an explosion.

Never run steam under pressure. It's not necessary for bending and turns your setup into a potential bomb. If you live in a higher altitude, your boiler may require some pressure to raise the temperature to 212 degrees; if this is the case, seek the advice of a professional boilermaker. ■

By far the most common failure, this breakage is the result of too much tension, or stretching, on the outer surface. More end pressure is required to make this bend.

WHY DO BENDS FAIL?

The real question is probably not why bends fail, but how in the world do they ever work? Still, taking a close look at the common reasons for failure will help with troubleshooting later on.

The most common failure results from the wood fibers on the outside of the curve being stretched to the breaking point. (Remember that in order to bend, the wood on the convex side has to stretch, the wood on the concave side has to compress, or a combination of these actions has to take place.) The tension generated from stretching pulls the wood fibers apart: The lignin, already softened by heat, simply lacks the strength to keep the fibers together. Since wood is more easily stretched than compressed, it simply tears itself apart on the outside surface in the process of bending.

Often a lack of heat in the steam box is at the root of this problem. The piece must be heated to a minimum of 200 degrees through to its core to become plastic enough to bend. If it is heated only on its outer surface or to an inadequate temperature, the chance of a successful bend diminishes rapidly. This is why it generally takes one hour per inch of thickness.

Even though the wood might be thoroughly heated, the bend will still almost certainly fail if the radius is small and the piece is thick. This is where the compression strap saves the day.

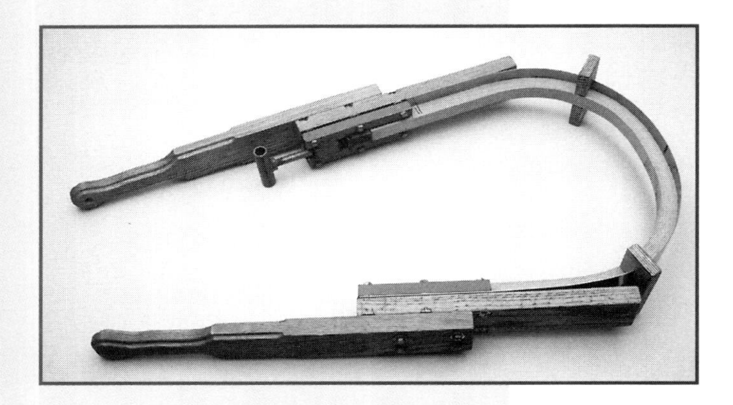

The key to bending solid wood, a compression strap limits the amount of stretching. The screw mechanism makes it possible to adjust the pressure on the ends of the bent piece.

The role of a compression strap

Though people have been bending wood for centuries, a fairly recent technological development—the compression strap—revolutionized steam-bending. The strap bends along with the wood, forcing the wood to compress along the convex surface rather than stretch along its concave surface. The strap accomplishes this feat with the use of end stops attached to a length of sheet metal. (See pg. 157 for info on building straps.)

During the last century, use of the compression

strap was developed to such a highly mechanized level that the high-volume production of steam-bent parts became very common. Michael Thonet, for example, the famous designer and manufacturer of ornate bent-wood chairs, was instrumental in devising new ways to make even complicated bends on a production basis. His chairs use bends that curve in several planes and still represent some of the best examples of complicated bent work imaginable. These chairs are still in production, having been produced by the hundreds of thousands over a span of over 100 years.

This commercially available strap offers an adjustable end stop and prefabricated strap material.

The key to bending with compression is end pressure. The correct end pressure will allow the piece to stretch no more than about ten percent of its length. The rest of the bend comes as a result of compression. Too much compression, however, can result in failure, as the wood fibers crush and the grain distorts by literally folding in on itself (pg. 139). Even if the piece didn't break, the bend would be no more successful than if it were to fail because of tension.

How much end pressure is needed? The amount for a given bend is hard to predict, but the rule of thumb on end pressure is that more is almost always better than less. If you experience bending failure due to excess end pressure, you will have managed to control a very powerful force. This confirms your strap is substantial enough to do the job, and it is then a simple matter to back off on the pressure the next time.

Compression sufficient to overcome the natural stretching of the piece being bent also works to control a host of other factors that contribute to bending failure, and generally makes the bending process much more forgiving. Factors that would have meant near certain failure, such as heavy runout on the grain of the wood, can usually be neutralized by compression.

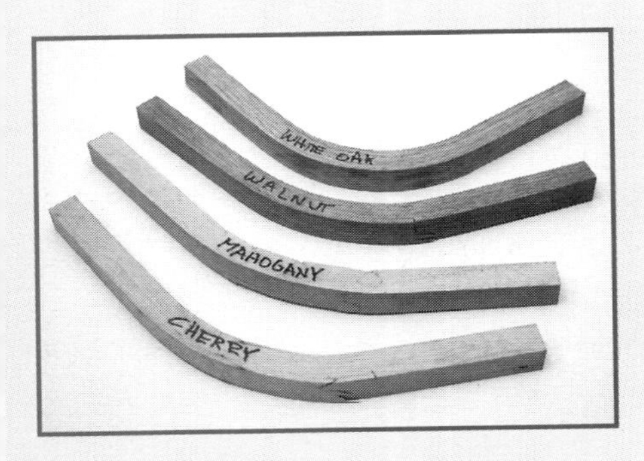

These sample bends were done at the same time using the same setup and kiln-dried wood (from top to bottom, oak, walnut, mahogany, and cherry). Most species of hardwood don't bend as readily as oak, which in this sampling was the only successful bend.

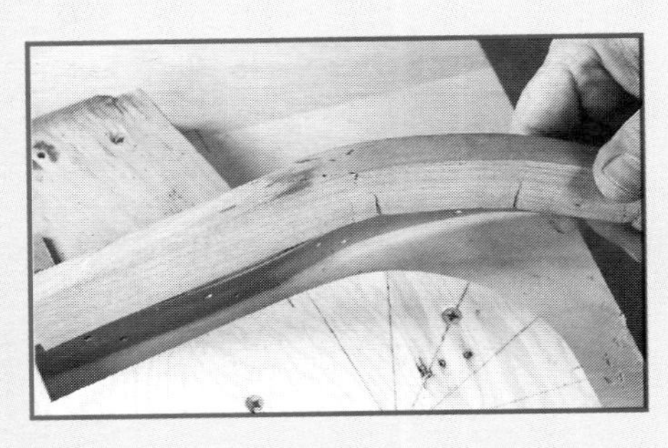

Cherry is always a challenge. It tends to fold up on itself from compression. But without compression, it fails in tension or the outside splits. So it's not a matter of failing either from too much compression or not enough. Cherry usually fails from both compression and tension.

BENDING VARIABLES

The possible variables with steam-bending are almost endless. Listed are the most important ones, but surely you will discover others. It will be impossible to control them all no matter what you do. All you can ask is to control enough of them so that the process eventually works.

Exotic wood species

For purposes of discussion, let's define "exotic" as any wood that is not oak. While any species can be bent to some degree, let's just say that some are more difficult than others. The difficulty or ease with which a particular wood is bent has to do with cell structure and a lot of other variables. There is a good reason, however, why boat frames are usually oak and snow shoes are usually ash instead of padauk or purpleheart. You will, of course, probably try to make that toboggan out of teak or something, but make sure that your setup works with oak first. Possibly, the lamination-bending process, discussed in Part 3, is a better choice if your heart is set on using something besides the best benders, which are oak, beech, ash, and hickory.

Moisture Content

It amuses me to read, as I often do, that only green wood freshly split from a log will bend. This is far from the truth. True, green wood is easier to bend than kiln-dried wood, since the high moisture content allows the core of the wood to heat quickly and thoroughly. If the wood has been split, as most green wood is, rather than sawn, it becomes even easier to bend. There's never a problem with grain runout when using split green wood.

On the downside, the more moisture the wood has when it is bent, the more it will shrink and check as it dries and cools and the less stable it will be dimensionally. Shrinkage is an ongoing and inevitable problem with green wood. It is possible to work around this characteristic, but for normal woodworking, where stability of the material is a strong consideration, there is a lot to be said for using kiln-dried lumber for bending.

Bending kiln-dried lumber

Kiln-dried lumber is harder to heat and harder to bend than green lumber. Without a compression strap, severe bends are all but impossible. Having said this, every piece of wood I've ever bent has been kiln-dried. It works just fine as long as you realize you're more likely to need a compression strap and more bending force to achieve success. As a bonus, kiln-dried wood stays stable after cooling and can be used in a furniture piece as soon as it cools.

It may surprise you to learn that soaking the wood even for a day or two has little effect on its actual moisture content. If a piece of kiln-dried wood is soaked in water, some water is absorbed, of course, but less than intuition might suggest. In any event, to the degree that water is absorbed, the wood will heat somewhat more readily, but beyond that there is little advantage to soaking.

There is much to be said for harvesting your own wood. Brian Boggs, for example, a Kentucky chair builder, often cuts, splits, and cures his own lumber. By doing so, when he steam-bends a piece, he has already controlled a number of variables. He selected a straight-grained tree. He cut a section out of the tree close to the ground where the wood is the toughest. He split the tree as opposed to having it sawn by a mill so that the grain runs parallel to the length of the piece and there's no runout. He probably bent it green or air-dried the lumber so that it contains just the moisture content he thinks is right for bending.

For the rest of us, especially those of us who live in urban areas, the lumberyard is where we do our wood harvesting. While we have much less control over the material than woodworkers like Brian Boggs, that obviously doesn't have to stop us from making successful bends. It just means we have to be extra-selective at the lumberyard.

Go after wood with the straightest grain possible. The grain on surfaced lumber is the easiest to follow, but the same technique may be used on rough lumber. Hold up a board to the light so that the reflection highlights the grain pattern. You will readily see if the grain runs parallel to the board or if it runs off the edge. While you are looking at the board, check it for flaws such as knots. If used in bending, these flaws can dramatically decrease the chance of a successful session.

TRY OTHER SPECIES IF YOU MUST

Once your setup is working reliably and consistently with oak, substitute a piece of mahogany or another wood considered to be more difficult to bend. What you'll notice right away is that mahogany, for example, crushes quite easily on its inside face. The fibers that compress so readily in oak simply don't act the same with mahogany. Try the same experiment with hard maple. Chances are that you will have both compression failure (as with the mahogany) and tension failure due to excessive stretching. These woods are prime examples of why it is important to set up your process using oak. Just to make sure, after you beat your brains out trying to bend some maple or other more difficult wood, heat up a piece of oak and see how effortlessly it bends using exactly the same technique.

For good bending characteristics, beech is a close second to oak. Much of the bent-wood furniture produced in Europe and elsewhere is made from beech. It is more expensive and harder to find in this country than oak, but it bends beautifully and its finer grain gives furniture a more finished look.

Ash is typically used for a variety of projects, especially for sporting gear such as snowshoes and fishing nets. It is light and strong for its weight. It bends almost as easily as oak, but it is harder to find straight-grained ash than it is to find oak.

Hickory is somewhat difficult to find, but bends about as easily as oak. Cherry will bend but only with great difficulty and careful adjustment of end pressure during the bend; a good reason to use an adjustable compression strap. It tends to compress in pockets rather than evenly. Teak, walnut, mahogany, and softwoods in general bend poorly. They are not good candidates for severe bending either with or without a strap. They fail in tension and compression far too easily. Even with a compression strap, more severe bends are almost out of the question. Lamination-bending is a better choice for curving these materials. ■

CONTROLLING VARIABLES

It's impossible to control everything, but you can still bend successfully without controlling every last variable. The important thing to remember is that there is part science, part art, and part luck involved in bending. The more you manage to control the bending process, the better your luck will be.

As you begin your journey into the world of bending, keep a clipboard or notebook handy to write down the various times, materials, successes, and failures. It will be a good reference later on. Without writing information down, it is all too easy to forget exactly when you loaded the box or when the temperature hit the 200-degrees minimum mark.

For example, you might have noticed that your box would hold maximum temperature with the heat turned down to mid-range once it has heated up thoroughly. This saves on energy and lets you know that you have the ability to use a somewhat larger steam box if you choose. If you go a few months between steaming sessions, this is something you might forget.

If you notice that after a couple of hours the water level gets quite low, a reminder to add water after about an hour might save you a great deal of trouble.

If you had trouble bending something but eventually got it to work, try to analyze what went wrong and what you did to make the process work. Your discovery may well lead to a breakthrough. Having a record of your attempts, successful and otherwise, might at least save the embarrassment of making the same mistake six months from now. ■

Overbending and springback

One thing is certain with steam-bending—some springback will occur when the piece is taken off the form. Just how much springback to allow for depends on each project. While the shape of the form must allow for some overbend, just how much overbend is one of those variables that's difficult to quantify. Many factors affect how a bend will turn out, including the radius of the bend, the thickness of the piece, how long the wood has been up to temperature, how long the piece sets in the form, and factors such as flaws in the wood. The speed with which you bend the piece on the form has a big influence on springback. The faster you bend, the less springback there will be. Likewise, the greater the end pressure, the greater the compression and the less springback will occur. In addition, milling the wood after it's been cooled may cause the bend to straighten out somewhat, which is another reason to allow for ample overbend.

A good rule of thumb is to overbend by about 20% to 30%. This should allow the piece to relax enough to settle in at the desired final shape. If the bend winds up too severe, it will be easy to straighten it a bit. By contrast, if the bend isn't curved enough the first time, it will be impossible to bend it further. Therefore, plan to overbend rather than underbend.

I commonly build my bending form so that it overbends the piece about 20% to 30%. I then let the piece relax just a bit by building my drying forms with 10% to 15% overbend. As the piece comes out of the box still hot, it relaxes a bit. When I clamp it to the bending form, it cools and stabilizes. After about 24 hours, the piece will relax a bit more when it comes off the drying form. With very little trial and error, it's entirely possible to build a set of bending and drying forms that will produce predictable and constant bends.

THE BENDING BLANKS

Always cut a few extra pieces, just in case. It's good to have a few pieces to warm up on, and spares in case something goes wrong.

Keep in mind that the piece will swell a bit during the steaming, so plan to give it a little clearance if it is a tight fit in the strap. (See more on this in the compression strap section.)

Grain Orientation in a square blank

Is there one grain orientation that bends better than another? Does rift-sawn wood, for example, bend better than wood with vertical grain? If the growth rings run flat around the radius, will the piece bend better than if the rings are on edge?

If the piece is square, it may be bent in one of two ways. To decide, check for grain runout (pg. 140) on each face. Choose the direction of bending that will place the face with the least runout on the side of the bend, not on the convex or concave face. If you mark the piece with a felt pen, it will save a few precious seconds during the bend.

Intuitively, it seems like the orientation of the end grain should effect the ability of the wood to bend. But having bent hundreds of pieces, I have found that there is no single grain orientation that bends better than another. Rift, flat, vertical—it makes no difference as long as the piece is without major flaws and grain runout, is heated correctly, and is bent with sufficient compression around a radius that isn't too tight.

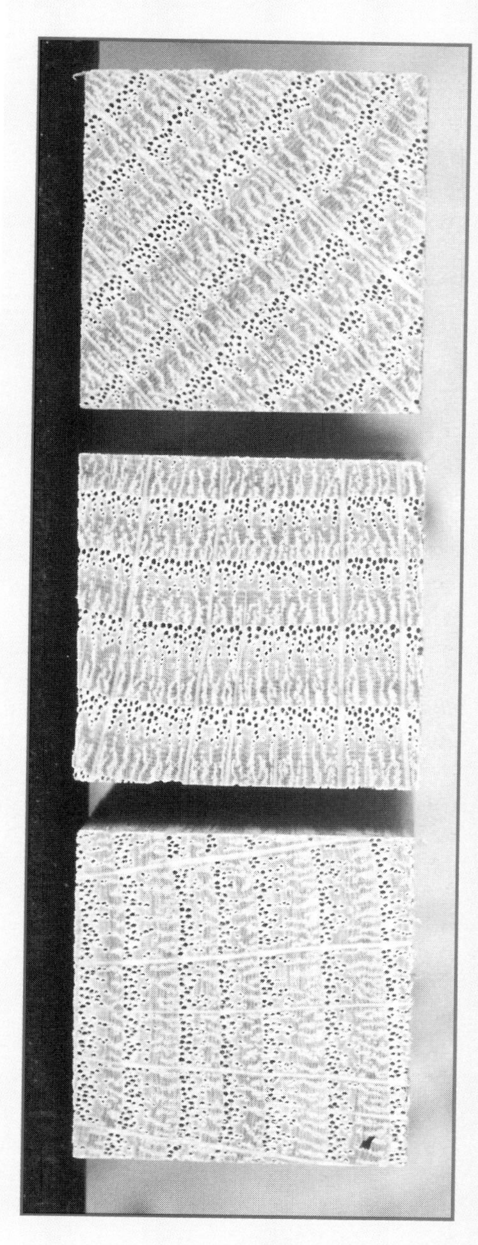

Surprisingly, it makes little difference which way a blank is oriented—whether rift, vertical, or flat—in relation to the bend. Runout is a much more important factor.

Milling the blanks for bending

The pieces to be bent must be accurately cut so that they fit your setup. Bending blanks should always be larger and longer than the finished piece. A certain amount of material should be left for the final trimming after bending is finished.

A well-made table saw cut is usually smooth enough for bending purposes, but the smoother the surface of the piece, the better. The piece will need to be sanded after it is bent anyway, but planing or jointing the blank prior to bending may be easier while the piece is straight.

Can you mill the profile prior to bending? This depends on the profile, to a large extent. Can you bend crown molding? Probably not. Can you mill the chair hoop round before you bend? Yes.

If we take the example of a chair hoop or a Windsor-style back or continuous arm, it is often feasible to mill the shape of the piece first and then bend it. If you first mill a piece, be mindful that the steam will at the very least raise the grain of the piece no matter how carefully you sand it prior to steaming. At worst, especially if you are using a metal mold or a steel compression strap, contact between wet, hot wood and steel almost always results in black stains on the piece. These stains typically penetrate deeply into the material.

Another factor to consider is that due to the stresses imposed on a piece when it's bent, some distortion may take place. A circular section of a chair part, for example, will wind up slightly oval after it is bent. The wood is mashed between the compression strap and the form and the round shape is crushed into an oval. These are among the reasons I almost always mill the profile after I bend, not before.

This pressure cooker on my ancient camp stove was my first attempt at making steam. I used it to "heat" a box that was 12 feet long and about a foot square. The boiler probably raised the temperature in the box about 5 degrees.

STEAM-BOX EQUIPMENT

Whatever you cobble together to make steam will invariably have something about it that could be improved. From the dozen or so rigs I've put together over the years, all I can say is that there is little doubt that my current setup won't be the last. The setup need not be elaborate as long as it's safe. The point is that many of the materials can be purchased at yard sales and thrift shops.

This electric steam kettle puts out an amazing amount of steam, easily heating a fairly large steam box (in this case a foot square by 3 feet long). Water is added throughout the steaming process with a funnel.

The boiler

Visualize an elderly, gas-powered Coleman stove with a pressure cooker on the burner spitting steam out of a hole about the size of a pencil lead. A piece of rubber tubing directs the steam into a box that is about 1 foot square and 12 feet long. The burner is going full blast. How hot is it likely to get in the box? Not very. I am embarrassed to admit that I just described my first attempt at steam-bending. There was simply not enough horsepower in the boiler to heat the box. The moral of this story is that while the boiler need not be high tech, it must be able to produce enough steam to heat the steam box. I mean *really* heat it—to the point where the steam box is at the required consistent 200 degrees minimum. Remember, it is mostly the heat that softens the wood, not the moisture. If the box is too cold, the boiler is not big enough. It's about that simple. Remember, you don't have to worry about overheating the steam box. At sea level, unless the box is under pressure, no matter how much steam goes in the temperature will never exceed 212 degrees.

For fuel to heat the boiler, gas and electricity work well. Electric heat is safe and it's easy to make steam with a variety of appliances ranging from old deep fryers to wallpaper removers and tea kettles. In good weather, an exterior gas barbecue is surprisingly effective. You can rig up a water container that sits right on the grill area. Either use hoses that connect the boiler to the steam box or set the box right on top of the boiler. Of course you will have to protect the hoses and the box from open flame.

Whatever form your boiler takes, it must be able to make steam continuously for several hours. This means that you need to have an adequate supply of fuel (it must last as long as the steaming process), and some provision for both checking the water level and safely adding water.

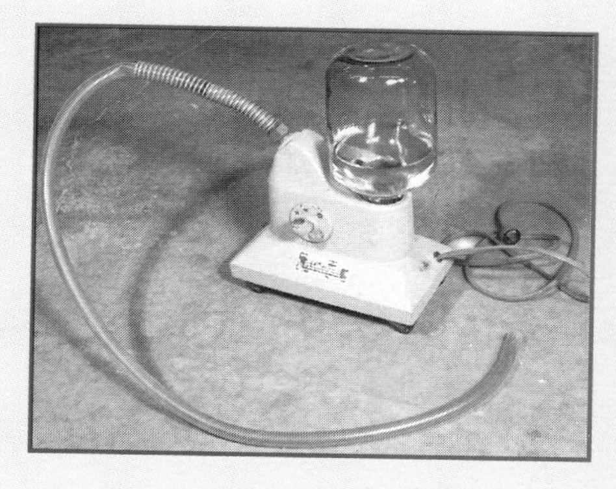

This device, used for stripping wallpaper, puts out steam but in fairly small quantities. Its visible water supply makes it easy to make sure there's always enough liquid.

This tank boiler has two propane burners and will put out serious steam—enough steam to thoroughly heat a box that measures up to 1 foot square by 12 feet long.

Getting steam to the box

The steam typically gets from the boiler to the box through some sort of hose arrangement. One of my steamers, built out of a borrowed electric deep fryer, had a steam box right on top of the boiler itself. This not only acted to minimize the heat loss between the boiler and the box, but also recirculated the water so I didn't have to top off the container as often. But most boilers will have to transfer the steam to a separate steam box through hoses. The larger the hoses, the better—a radiator hose is just about the right size. Insulating the hose will cut heat loss to a minimum. Pipe insulation works fine.

If the box is much over 3 feet long, either manifold the steam into the box in several places or consider using more than one boiler. This way the steam will not have to travel farther to one part of the box than to another. The idea is to create uniform heat throughout the box.

This boiler started out as a borrowed deep fryer. Its steam box is shown on the next page.

Gauging the water level

The water level will need to be monitored in some way. In a

Wire mesh above the boiling water in the fryer keeps the pieces suspended in the steam. It's galvanized, which keeps it from staining the wood black.

This box fits over the deep-fryer boiler. Much of the condensed steam automatically returns to the boiler, reducing the need to top off the boiler with water.

perfect world you'd have a sight gauge, but in an imperfect world even a dip stick will do. Water will need to be added frequently during the process, so your rig must include a valve or opening of some sort. Since it's important to keep the water boiling, add water sparingly and frequently instead of in large quantity all at one time. This will ensure you don't cool down the water and the box. If possible, add preheated boiling water.

Building a steam box

The steam box is one of the most important components of the setup. It must allow quick access for loading and unloading the steamed pieces in the least possible time. It must allow the steam to circulate evenly around the pieces. It need not be a work of art, but it must be made strongly enough to withstand the rigors of steaming. Most important, it must stay hot. The smaller the box is, the less energy it will take to heat and the smaller the boiler you'll be able to use. Make the box only as large as necessary to contain the pieces you're bending. If bending something 4 feet long, the box needs to be just a little longer than 4 feet long inside. (Depending on design, sometimes it's possible to heat only a portion of a piece and let the ends of the piece stick out of the box.)

Exterior-grade plywood is ideal for building steam boxes, but just about anything around the shop will do. Keep in mind that if you use solid lumber for the box, chances are that it will warp and distort a great deal when heated. The door may jam right at the worst possible time. This is why the more stable plywood is the preferred material.

When building, don't make the box airtight—you want the steam to circulate. Make sure the door will operate when the box is hot and wet. If you want shelves, use dowels, which don't block the circulation of steam. Add a drain to handle condensation. Since hot runoff water from the box may stain concrete and kill grass, it's a good idea to drain the box into a bucket so the water can be disposed of without damage to floors or landscaping. For drainage, I usually build my boxes with legs and mount them on a slant so the condensed water will run out of one of the ends. (Adding legs to the box also makes it easier to insulate as well as easier to load and unload.) And by placing a cookie sheet or other container under the drip, you can contain much of the mess.

Avoid using metal in the steam box. During steaming, the inside of the box will become very wet. If the box is horizontal, condensation will form on the ceiling of the box and drip on your wood. If the condensation comes in contact with steel nails or screws, and your wood is oak, expect dark splotches. I recommend that you assemble the box with drywall screws and polyurethane glue. Once the glue dries, remove the screws. Polyurethane glue holds remarkably well around the hot and wet environment of a steam box.

PVC or plastic pipe seems like a logical choice for a steam box, but in reality it doesn't work all that well. As the pipe heats, it softens. If you have the pipe on a pair of sawhorses with your bending pieces inside, the pipe will start to sag in the middle as soon as it heats up. Soon the bending pieces will be the only things keeping your steam box from folding up. One way around this is to build a V-shaped trough to support the pipe along its entire length.

Steel pipe, another seemingly logical choice for a steam box, should be avoided. If you have ever seen the black mark a steel bar clamp

Exterior plywood works well for steam-box building. The box should allow for some circulation of steam and should not be airtight.

PVC works okay as a steam box as long as you realize that as soon as it heats, the plastic will droop. It must be supported along its entire length in a trough supported by sawhorses.

In a vertical steam box, I use dowels to separate the pieces. Otherwise, they'll lay in a clump and the steam won't be able to circulate around them.

Dowels suspend the parts inside this steam box. This way steam can circulate all around the wood, heating it evenly.

makes if it lays in glue overnight, you'll understand why. A better, but much more expensive, choice for a permanent installation is stainless steel or aluminum pipe.

This simple form for building a chair crest rail is easily made from scrap plywood. Its radius must be smooth, providing a solid mold for the bend.

FORMS

The form, or mold, around which a piece is bent will be different for each project. Whatever its shape, the form must be strongly made, must provide for quick clamping, must take overbend into account, and must be as smooth and fair as possible because any bumps or hollows will transfer to the new bent shape.

I assemble forms using glue and screws. To hold the form to the bench during bending, I typically clamp it in a vise. I attach some sort of appendage to the form that will fit the vise or provide clamping that won't interfere with the bend. In the case of the crest rail (opposite page), it was easier to bend the piece by standing on it and using my body weight, so I simply placed the form on the floor. When it comes time to bend the pieces, I like to use bar clamps, which are quick to apply. Remember, you have but a few seconds to make the bend before the wood starts to stiffen. Cut pockets in the form for the clamps anywhere you want.

A cleat bolted on one end of the chair-crest-rail form speeds the process by allowing the easy insertion of an end of the steamed piece.

I prefer exterior-grade plywood for making a bending form. Particleboard will work, but plywood is stronger and less apt to distort under pressure. Plywood is also better able to withstand repeated clamping operations; particleboard tends to flake away after only just a few uses.

Cutting out the layers of the form

I like to use 3/4-inch thick plywood for making bending forms, since it's easy to handle and inexpensive to purchase. The layers stack on top of one another to create the desired dimension. Rout the first layer of plywood using a batten as a guide (discussed in Part 2), which will give you a pattern for the rest of the layers. Then flush-cut the other layers until you have enough thickness for your form. Glue together the layers, using screws to hold the assembly until the glue dries. Keep track of the

Bending on a form

After inserting an end into the cleat, it's a simple matter to bend the wood. Notice the gloves—the wood is fresh out of the steambox.

A bit of foot pressure helps the bend along.

After the piece is bent about halfway, take it off the form, quickly turn it around, and finish the bend on the other end. This pre-bending helps the last 6 to 12 inches—the hardest part of this crest rail to bend—to make it through the process intact.

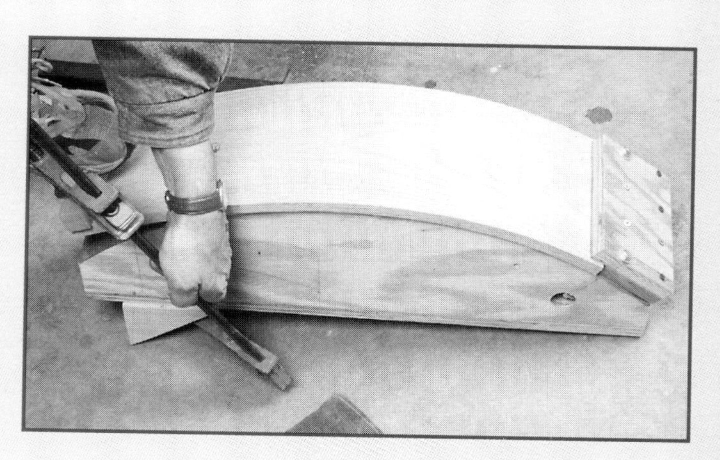

Clamps pull the crest rail into the form for the last and usually the hardest bit of bending.

The block on the end of the part keeps it from splitting by spreading the clamp pressure out over the entire width of the part. In addition, the block keeps the end of the part flat.

These drying forms keep the piece bent in the desired shape while it cools.

locations of the screws, because you'll be drilling holes in the form after it is assembled.

Sand the outside surface of the form until it is smooth and the curve is fair. The form must be as smooth as possible. Any lumps or hollows will telegraph right into the piece being bent. If the form is uneven, the piece will not have a fair curve.

The drying (cooling) form

I've discussed the drying form a little on pg. 148. A drying form holds bent pieces while they cool and stabilize. Using a drying form speeds up the bending process by freeing up the bending form and compression strap almost as soon as the piece is bent. Most sources say to leave the part on the bending form at least until it cools, but if you use a compression strap and a drying form, leaving the piece in the form for about five minutes is plenty. Once compression of the inner face takes place, the stress in the piece just about disappears. There is no "twang," as if the piece had flown off of the form. It springs back a bit when the clamps are loosened, but not a lot. Another advantage is that a drying form can go a long way toward controlling springback. The drying form need not be elaborate, but should keep matching pieces bent the same amount while they cool.

A screw mechanism like this makes it possible to adjust the pressure on the ends of the bent piece. As the piece bends, tighten the screw just enough to limit stretching.

TYPES OF COMPRESSION STRAPS

In bending, a compression strap is often the single difference between success and failure. Regardless of the severity of the bend, a compression strap will control more variables than any other factor. While bending is possible without a strap, as the severity of a bend increases, so does the need for a strap.

There are basically two kinds of compression straps—those that are adjustable (so the amount of end pressure can be varied), and those with fixed stops (where the pressure is set before the bend is begun). The great advantage of an adjustable strap is that you can increase or decrease the end pressure as needed during the bend. If the strap incorporates a screw device for tensioning, the pressure may be gradually released during the course of the bend, adding a measure of control to the process. When bending tricky woods such as cherry, it's sometimes possible to save a bend that would otherwise have too much compression by backing off the adjustment just a bit as the bend is drawn into the form. Moreover, the adjustment is much faster and easier than trying to shim a hot blank to minimize the amount of play in a fixed-stop strap. I've tried a number of ways, including wedges, but I invariably come back to either solid shims or my first choice—an adjustable strap.

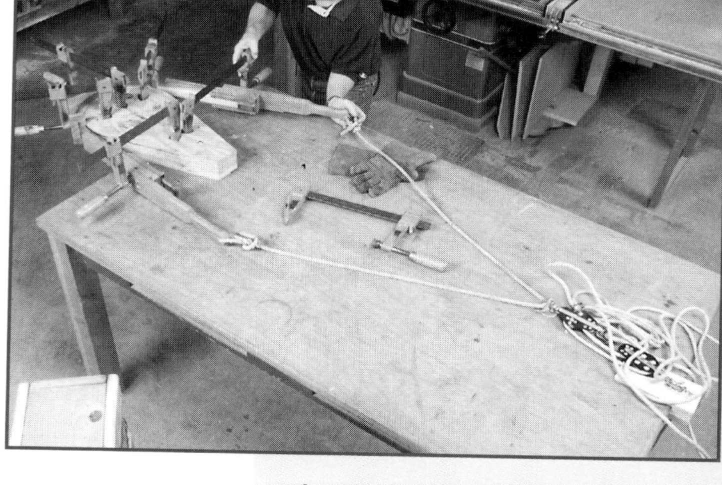

If working alone with a compression strap, as shown here, it helps to hook up a cable puller or a block and tackle.

If you have some metalworking ability, try making your own straps, as I do. Lacking that expertise, commercially manufactured straps work very well.

Building a strap

The length and width of a strap, like that of a steam box, will depend on the size of the part being bent. It is likely that during your steam-bending career you will make a variety of straps. A 4-foot piece needs a strap 4 feet long. A strap designed for a 1-inch square will not fit a 2-inch width. While you can design a multi-use strap, you may find that it isn't that hard to build multiple straps once you get used to working with metal.

The end stops on this scaled-down strap won't last forever, but they're plenty strong enough to bend a few sets of chair parts.

A good hacksaw, a set of regular twist drills, and a center punch are about all you need to build a simple strap. Use galvanized steel or stainless steel for the strap. This material will prevent the staining of the wood that invariably happens when damp hot wood contacts steel. Spring steel would be the best to use were it not for the fact that it is difficult to cut and drill. Regular old mild steel works fine.

Look for a steel supplier near your town that sells remnants by the pound. You'll usually find what you need to build straps there for just a few dollars. If you need a part that you cannot cut with tools you have, a steel supplier can usually cut it for you. If you don't have a metal supplier, you can use the metal framing hardware available at your lumberyard or home center. Try the galvanized coiled strapping material used for diagonal bracing when framing. This material is flexible and comes in different widths and thicknesses.

The stops are built from home-center parts. The metal strap is diagonal tie strap for framing. The rest is scrap wood, nuts, and bolts.

By contrast, this angle-iron end stop simply wouldn't be up to the task of limiting the forces trying to stretch the wood as it was bent.

A sample strap

I'll describe building a compression strap for bending a Windsor chair back about 1 inch square, 42 inches overall, and bent through about 180 degrees.

First cut some blanks to the size you'll be bending; size the strap accordingly. Gather your tools and materials, a box of coiled galvanized strapping, six 1/4-inch carriage bolts, some scraps of hardwood, and a set of twist drills.

I always draw pieces full scale so I know the shapes and sizes before I cut. Cut out the wood pieces first. The handles extend along the strap to act as backer blocks, which prevent the end stops from rotating.

After you cut out and drill the wood pieces, use the holes to mark the location of the holes in the strapping material. Punch the centers of the holes with a center punch, then drill the holes in the metal strapping. Twist drills will drill the metal quite easily. Apply a lot of pressure while drilling, but ease up just when the drill begins to punch through the side. Don't cut the strap to length quite yet.

Assemble one end of the strap. Then insert a bending blank. Clamp the other end stop to the strap and mark where the holes should go and the length of the strap. Now you can hacksaw it to length and drill the holes.

The most important aspect of building a compression strap is making sure that the end blocks are securely fastened. To be effective, the end blocks must be built to withstand thousands of pounds of end pressure. Use the largest diameter bolts possible. This will strengthen the entire assembly.

Assemble the other stop and handle, and ensure the blank fits into the strap well enough so that you can install it in a hurry. I round all

Compression strap and handles

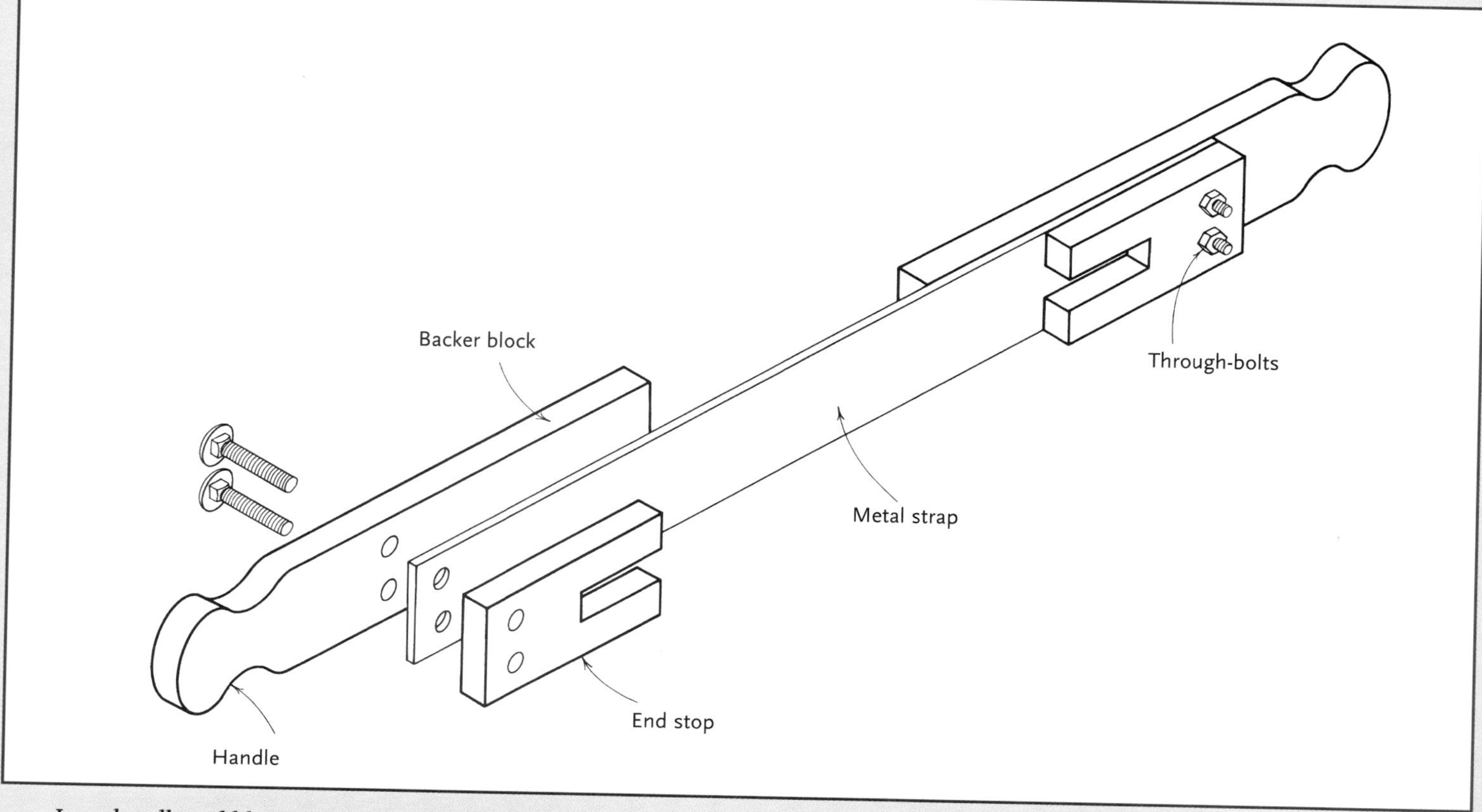

Long handles add leverage for bending. The handles should be smooth and provide a good grip. Extend the handles up along the strap to act as backer blocks. Backer blocks counteract the tendency of the end stops to pivot out of alignment, allowing the bent piece to pop right out of the strap.

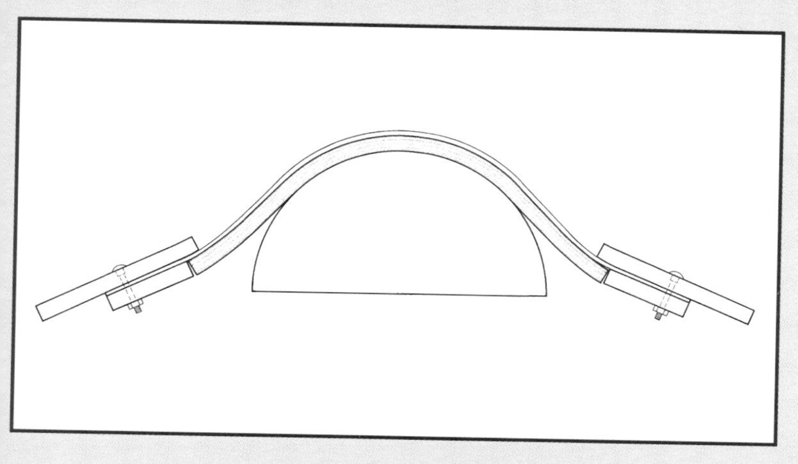

Backer blocks of insufficient length won't prevent the stops from rotating out and off the ends of the piece being bent.

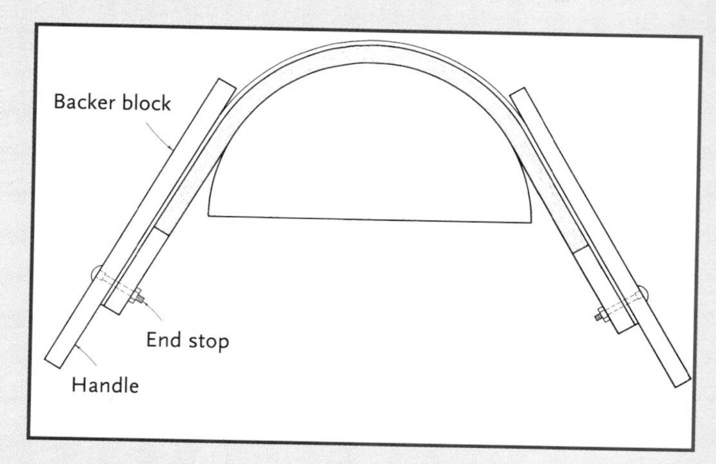

Extended handles work as long backer blocks to promote a successful bend.

Bending on a form with a compression strap

To apply the maximum end pressure to the piece, shims go between the end stop on the compression strap and the end of the part. An adjustable end stop is easier, but shims work fine if you have only a few boards to bend.

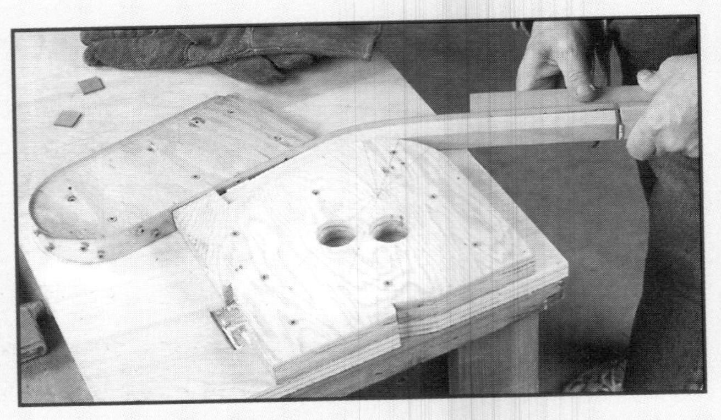

As the bend begins, end pressure builds up, forcing the part to compress and not allowing it to stretch the way it naturally would absent the strap. Securing the jig to the bench is a must.

At this point in the bend, considerable force is necessary to get the part to bend around the form. You may have to lean into the bend with your hip. A longer handle would give more leverage to this strap.

The bend is almost as far as it can go with hand pressure. Notice the jig is shaped so the part is bent beyond 90 degrees to allow for springback. As soon as the part comes off the jig, it will go onto a drying form.

A large bar clamp pulls the part into the form for the last bit of bending. The handle serves several purposes—it gives leverage for bending, prevents the stop from springing out from the end of the part, and acts as a clamping block to flatten the part where needed.

The crinkling, especially noticeable on the inside of the bent part, shows how compression takes place as the piece bends. Without this compression, bending is nearly impossible.

the sharp edges, especially on the handles. Tighten everything up and you're ready to bend.

If this strap seems overbuilt, you'll soon see that the opposite is true. After only a few bends, the strap will have stretched and distorted to the point where it will need to be replaced. That is also about the number of bends it will take for the advantages of adjustable stops to sink in. Meanwhile, you'll need a handful of shims the size of your bending blank. I like to use pieces of plastic laminate, which are readily available and don't easily crush.

HOT-PIPE BENDING

This technique is most frequently associated with musical-instrument making. The hourglass shape on the sides of a guitar, for example, is most commonly made by slowly bending the wood over a heated pipe. This method works fine for pieces measuring 1/4 inch or less in thickness.

As when using any other bending method, start with blanks that have been selected for straight grain, are free from blemishes, and are carefully surfaced. Use a drawing or pattern of the finished bend as a template against which to judge your work as you proceed.

Some theory

Although the heat is dry as opposed to moist, the principles of hot-pipe bending are the same as those of conventional steam-bending. The piece being bent has to be wet. When it is laid on the hot pipe, steam forms, which heats and softens the wood on the inside face. Because the inside face is softened by the heat of the pipe, the outside surface is under less tension. When the piece starts to bend, instead of the outside failing due to too much tension, the inside of the piece can compress because of the heat.

SAFETY TIPS FOR HOT-PIPE BENDING

Wear gloves and short sleeves. It's easy to be concentrating intently on the bend and, without realizing it, catch your long sleeve on fire. Please don't ask me how I know this. But do be mindful that the invisible flame from the torch is exhausting in your direction, hence the baffle. Although the baffle will help to direct the hot air away from you, you still need protection. Gloves are always a good idea. I often wear welding gloves that cover my forearms.

Use black, nongalvanized pipe, since galvanized pipe emits a toxic metallic cloud when it gets too hot.

The pipe can get extremely hot, so hot it can burn the plywood. Take care to get the pipe hot enough to bend the wood, but not so hot to scorch the plywood.

Finally, let the pipe cool thoroughly before you touch it or put it away. ■

For less than $20, you can make your own hot-pipe bending setup. All you need is a small propane torch, a 2-inch pipe nipple, some screws, and a scrap of plywood.

Building the hot-pipe setup

Any sort of small propane torch will work as the heat source. The size of the pipe depends on the radius you're trying to bend. The pipe nipple in the photos is 2 inches, but just about any size up to 3 inches would work fine. In addition, you'll need a scrap of plywood and a few screws.

I like to create a flat surface on top of the pipe to preheat the piece and also to form a smaller radius on each end for tighter bends. You can flatten the pipe with a large vise or with a sledgehammer. If the pipe is still reluctant to bend, heat it with the propane torch. When it's bent, be sure to cool it thoroughly.

Next, using a hacksaw or reciprocating saw with a metal-cutting blade, cut tabs on one end of the pipe so you can fasten it to the plywood. Heat the tabs, then bend them out to 90 degrees. Drill holes in each tab to accept the screws. Cut a hole in the plywood the same size as the opening in the pipe. Then screw the pipe to the plywood.

A baffle in the end of the pipe will both conserve fuel and help direct the hot air blowing out the end of the pipe. I cut this one with scissors out of an aluminum can. Stuck into the end of the pipe, it keeps the hot air somewhat confined in the pipe. I didn't use a threaded cap

Cut tabs on a pipe end, drill the tabs so you can insert screws through them, cut a hole in the plywood, then screw down the pipe.

This pipe baffle was cut with scissors out of an aluminum can. To use the hot-pipe setup, place the plywood in a vise, setting it so the torch can reach inside.

Flatten the pipe so it's slightly oval-shaped. This will give you a better surface on which to heat the material, and a smaller radius on two sides in case you wish to bend a smaller radius.

because I wanted to flatten the pipe and I didn't want to completely seal off the end. The flame would go out if I did.

To use the setup, place the plywood in a vise. Then light the torch and place it so that the flame is directed to the inside of the pipe. Don't use a very high flame, since you don't want to overheat the pipe. To test the temperature of the pipe, dribble a few drops of water onto it—if the water doesn't quickly boil, the pipe is too cool; if the water skitters off immediately, it's too hot. If the water sits on the pipe until it boils away, the pipe is just the right temperature for bending.

Moisten the side of the stick that will be against the pipe with a wet sponge. Do this every few seconds, keeping the stick wet on the inside of the bend. This both generates steam and prevents scorching.

Bending the wood

It's a good idea to soak the pieces in water for a little while prior to bending, especially if using kiln-dried wood. Then, using a sponge, continuously wet the side of the blank that will be against the pipe so that it doesn't dry out. This keeps the blank from scorching and provides moisture to generate steam within the blank.

To bend a small piece, lay it on the pipe and slowly rock it back and forth on the part of the blank where you want the bend. Apply light and steady pressure, and be patient. It may take several seconds until the piece heats enough to bend. Eventually, though, you'll feel the piece start to "give." Develop a feel for this by practicing with scrap. It's a matter of letting the piece bend very gradually, not forcing it. It'll bend naturally and easily, but only when it's ready. Remember to keep the inside of the bend wet, both to generate steam and keep the wood from charring.

Carefully rock the piece back and forth where you want the bend, using light and steady pressure.

Hold the blank against the pipe, applying steady pressure until it's bent like you want it. Then go a bit further. By bending the part a little more than you really need, you can rest assured that it's far enough even if the bend relaxes a bit more than you think it will. Remember, as in steam-bending, it's quite easy to unbend a piece a little, but try to bend it any further and it will almost always break.

Finally, hold the wood against the pipe for a few seconds more until the bend is set, then hold it in the shape you want until it cools.

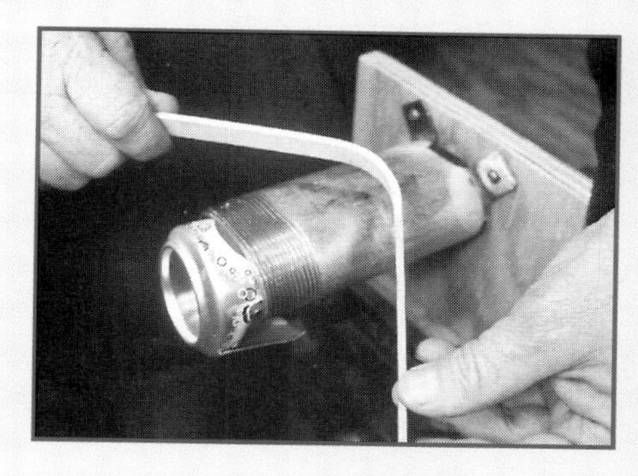

Shape the bend you want, bend the wood a bit further, then hold it against the pipe for a few seconds to set the bend. Hold the wood in its bent shape until it cools.

PART FIVE

Combining Techniques & Other Methods

My clients wanted their staircase to be simple and contemporary. It took just about every trick in the book to pull this off. This stair uses a combination of steam-bending, bent lamination, and curved milling. When I designed the stair and railing, I figured it would be challenging to build. I was right. The railing on the right is offset over a foot without any support in the middle. There are steel bands embedded in the wood to carry the load.

Combining Techniques & Other Methods

SOMETIMES ANY ONE METHOD of creating a curved piece simply will not do the job. At that point, it's worthwhile to consider combining curving methods or even to explore using one of the lesser known techniques such as coopering or kerf-bending.

Suppose, for example, that you have laminates of a certain thickness, but that the laminates are too thick to bend around the radius you need. An example of this is the pre-milled stock bending rail that suppliers provide for circular handrails. The core layers are approximately 1/4 inch thick and the outside molding contour is almost 3/8 inch thick. These thicknesses work fine for a staircase handrail that exceeds 5 feet in radius. But what about stairs 3 feet and under? This is what stairbuilders in the real world run into all the time. As discussed on the next page, one solution is to steam the laminates, bend them on the mold, let them cool, glue up the laminates, and bend them again.

Here's another example: Perhaps it's desirable to have the wood grain running along the curvature of certain furniture parts, but the shape of the parts must be extremely precise. Here you might first steam-bend the parts to approximate shape, let the parts stabilize, lay out the precise contour, and then mill the steam-bent parts to exact curvature.

Sometimes you might even find yourself employing all three primary methods of curving wood in the same piece. For example, the piece might be steam-bent in one spot, milled in another, and the whole thing lamination-bent. The handrail in the photos on these pages used steam-bending for the sharp bends at the beginning step and bent--

This railing transition is made up of nearly a hundred separate pieces of wood. The curved tread nosing was cut on a shaper using a wagon wheel-type pivot (pg. 63), with a 6-foot radius. The treads are milled. The risers and side skirtboards are bent laminations.

The section of the railing that bends down to the floor is steam-bent. The part that runs parallel to the stair is a bent lamination.

lamination for the large radius going up the stairs. The lower curves on the underside of the railing were milled to shape.

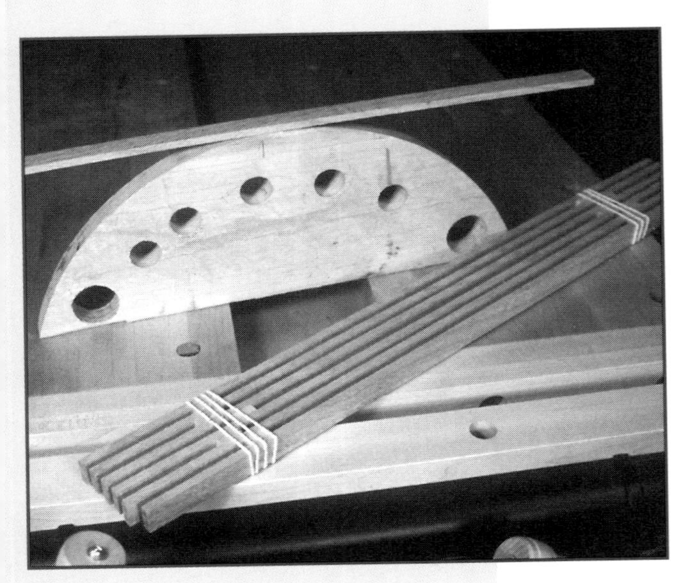

The bundle of strips in the foreground is ready to go into the steam box. By separating the layers with blocks, steam can circulate more freely to heat them evenly. Once hot, they easily bend around the form.

STEAM-BENT LAMINATES GLUED AND RE-BENT

Tight bends can be made using thick laminates by steaming the laminates and pre-bending them around the form prior to gluing and re-bending. Before steaming, separate the layers with 1/4-inch by 1/4-inch pieces of scrap so steam can circulate. Bind the bundle with string so it will stay together. Steam the laminates for just a short time (20 minutes or so); after steaming, remove the layers from the steam box and quickly cut the string and remove the spacers. Now carefully bend the entire bundle of laminates around the form. The next day, remove the bundle from the form and separate the layers so they'll dry thoroughly—an important step because moisture trapped between layers can cause gluing problems later. Hold the shape as much as possible during the drying phase, which should take about 24 hours. A version of the drying form used in Part 4 will hold the pieces just fine. Clamping the ends to a straight piece of wood or even tying the ends with rope will adequately hold the bend while the layers dry. The true radius will be bent when the layers are glued after drying.

The single layer of the bend on top of the form was obviously too thick to bend around such a tight radius without steaming first. By steaming the pieces and pre-bending them, thicker layers are feasible.

Gluing can be tricky, since curved pieces are awkward to handle. Spread the glue on both sides of each layer and bend once more, following the directions given in Part 3. Because the pieces are pre-bent to approximate shape, they should bend easily around the curve.

When you use this method, expect less springback than would ordinarily occur: Because the steaming relaxes the wood, it has less tendency to straighten out.

Is this a lot of trouble just to bend a handrail? Yes. But is there an easier or quicker way to get the job done? Not that I know of. If you think there is, it will probably involve—you guessed it—one or more of the other curve-making methods. Remember, there's no one perfect way to make a given part. What works for one project may not work for another.

And what seems reasonable for one part may be too much trouble to take for another.

Sometimes it's feasible to bend a single piece so you can cut two identical parts from it. This eliminates the problem with bending being somewhat inconsistent. Even if there is some variation from the original design, at least the two parts will match.

STEAM-BENT AND MILLED SOLID LUMBER

Steam-bending lacks the precision of milling and lamination-bending. The degree to which the piece relaxes after it is bent is difficult to predict no matter how carefully you control the process. More precise, consistent results are possible when the shape is milled after it is steam-bent. The rough shape of the steam-bent part may then be milled to precise shape.

Rocking-chair runners offer a great opportunity to use this combination of techniques. Runners must match exactly and must be precisely shaped. Unfortunately, they often are weaker than they should be because of grain runout. I would guess that this is the reason many, if not most, folks who build rocking chairs either cut the rockers out of solid stock or use bent lamination. I rarely see steamed rocking-chair rockers. Steam-bending runners to rough shape prior to milling them to final shape provides the best of both worlds: accuracy and elimination of grain runout.

Cut the blank to leave at least 1/2 inch for trimming after the bend is made. Bend the blank using a compression strap. If your setup is large enough, you can bend one piece that can later be sawn into two identical parts. After bending, let the blank stabilize on a drying form for several days. When the piece has cooled, lay out the pattern and mill as you would any solid lumber.

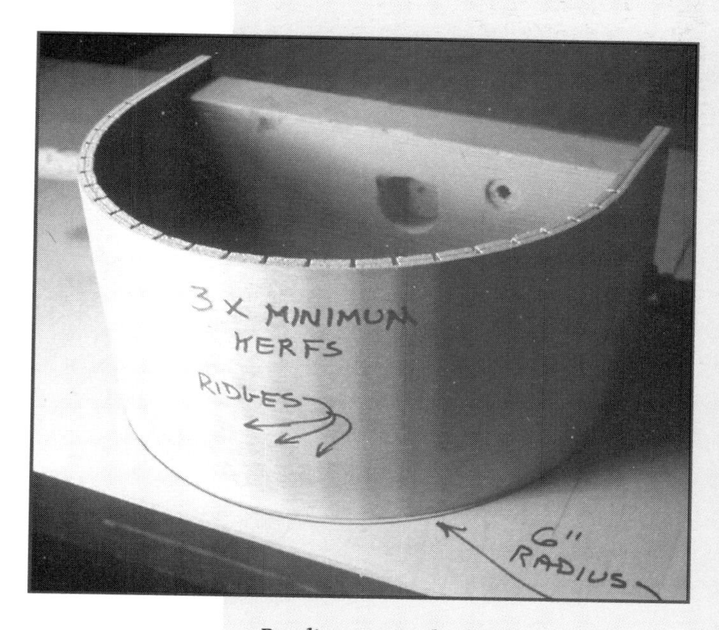

Bending a piece by kerfing the back surface is really nothing more than a form of bent lamination. The trouble with this method is that flat areas appear on the surface.

KERF-BENDING

A kerf is the slot left behind when a sawblade makes a cut in a piece of wood. If several cuts, or kerfs, are made along a board, this in effect makes the board thinner and more easily bent.

This bending technique is actually a variation of

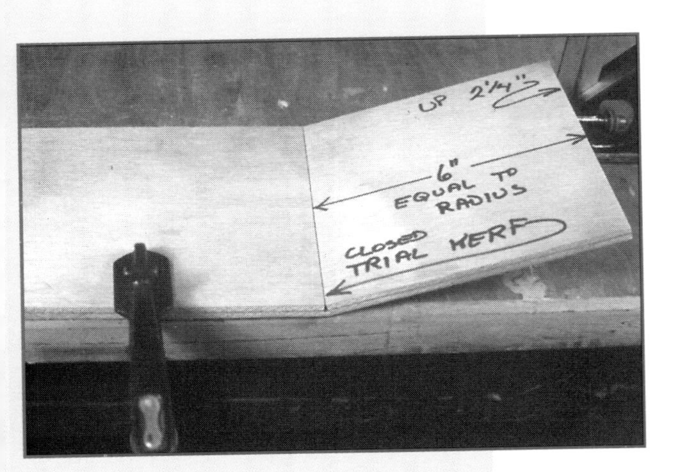

There is a simple way to see how far apart the kerfs must be. Make a single kerf and bend the piece as far as it'll go. Measure out the distance of the radius. The minimum spacing equals the height off the bench at that point.

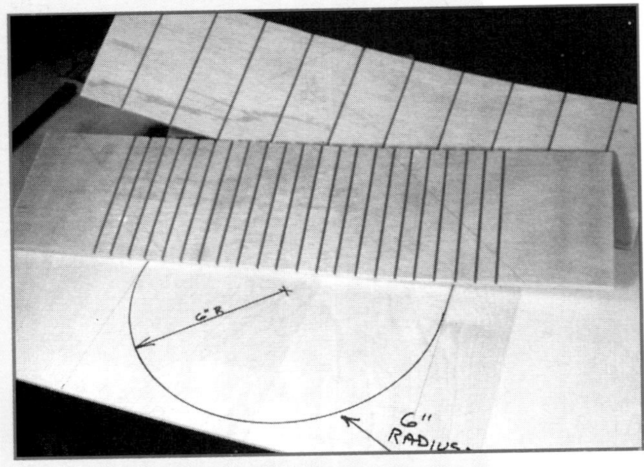

The closer the better for kerf-bending. The farther apart the kerfs, the more pronounced the cross-grain flat spots will be.

lamination-bending. The outer layer, which is left solid, is thin enough to be considered veneer. This veneer, cut sufficiently thin, will be flexible enough to allow the piece to bend. But there's a major downside when working this way. Once bent, the piece tends to fracture across the grain, eventually making an otherwise graceful curve unsightly.

Kerf-bending is a fairly easy way to make a curved substrate—an inner layer of structural material. Using the system I describe to figure the spacing and number of kerfs for a given radius, you can make a kerf-bent substrate that's just the right shape for the curve you are trying to construct. To solve the fracturing problem, a covering of veneer is a fast, neat, and attractive last step.

If you are kerfing plywood, set the depth of cut so that the last layer of plywood that runs the length of the piece is left intact. If using solid lumber, the kerfs must be deep enough to leave material thin enough to bend around the desired radius. Remember that the thicker the outer layer is, the more likely it is to crack. Yet the thinner the outer layer is, the weaker it will be. It's a balancing act.

Solid lumber is much more susceptible to fracturing than plywood. It's a good idea to make a prototype so you can judge if the process is going to work. A radial arm saw, table saw, or circular saw are all good choices for kerfing either plywood or solid wood.

The minimum distance between kerfs depends on several factors: the radius to be bent, the thickness of the stock, and the width of the kerf itself. There is an easy way to figure the spacing.

Make a trial kerf in a scrap piece of the same material you plan to bend. Measure off a distance equal to the radius along the board, starting at the kerf. Carefully mark this distance. Place the piece kerf side up on the bench. Now bend the piece until the kerf is almost but not quite closed. Holding the piece bent in this way, measure the distance between the bench top and the mark indicating the radius.

When a piece is kerf-bent, the bend takes place only where the wood is kerfed. Between the kerfs, where the board retains its original thickness, the board stays flat. When the bending takes place, there is a flat spot, then a sharp bend, then another flat spot, then another sharp bend, and so on. Over time, fracturing right where the kerfs are makes lines across the grain.

Sanding is rarely enough to hide the problem entirely, although it does help. If you sand enough to smooth the surface, you'll likely sand through the veneer if using plywood or through the outer layer of solid lumber. The thinner the outer layer of solid wood, the greater the chance that it will fracture. Eventually, you could sand all the way through the outer layer and expose the kerf itself. That's why it makes sense to use kerf-bending to create a substrate for veneer. Initially, this seems to defeat the purpose of kerfing, but if you plan to veneer before you start the project, the method works fast and looks great.

RULES OF THUMB WHEN KERF-BENDING.

The closer the kerfs are together, the easier the piece will bend.

The closer the kerfs are together, the less cross-grain fracturing there will be.

The closer the kerfs are together, the weaker the piece is.

Regardless of the radius, the kerf should be no more than about 1/8 inch from the outer surface of the bend.

The spacing should be consistent so the piece bends uniformly.

Where spacing is concerned, closer together is usually better than farther apart. ■

Substrates for bent veneer

Using a lower grade of lumber or plywood for the substrate, kerf the wood as described. Make a trial bend over a male form made of plywood or particleboard to make sure the kerfs are spaced correctly and are deep enough for the piece to bend easily.

Filling the kerfs with two-part auto-body filler makes a strong substrate curved to match the desired radius.

Next, mix up some auto-body filler or thickened epoxy. Using a plastic spreader or gloved hands, fill the kerfs as thoroughly as you can. Work fast because the mixture sets up quickly. When the kerfs are as full as you can get them, bend the piece over the form. Clamp or temporarily screw or nail it in place. As the part bends around the form, the filler will be forced out of the kerfs. It shouldn't take much clamping pressure to hold the piece in place while the filler sets up. Work more filler into the kerfs after the piece is bent, removing the excess as you go.

By gluing a thin veneer over the kerf-bent shape, the problem with the flats is eliminated.

When the filler cures, remove the piece from the form and clean off any excess. The filler stabilizes the curve, making the

With a simple setup on the band saw, it's easy to cut the veneer off a strip of hardwood veneer plywood.

kerf-bent piece almost as strong as it was before it was bent. You now have a curved substrate ready for veneering.

Veneer choices

For veneer, you have many choices. To list just a few, there is real wood veneer, paper-backed veneer, imitation veneer, and high-pressure laminate. You can also cut veneer yourself.

If you cut your own veneer, you can use either hardwood veneer plywood or solid lumber. Hardwood veneer plywood is a particularly good choice for this application. There is a way to cut a thin, flexible layer off a piece of plywood, keeping the veneer intact and using the first layer of plywood as a backing.

With typical hardwood veneer plywood, the outer layer of veneer goes lengthwise, or end to end, along the full length of the sheet. The layer of plywood just below the veneer goes across the sheet, or side to side. If you cut away the outer layer of plywood from the rest of the piece, you'll have a flexible piece of material that also has a good deal of thickness and strength—perfect for gluing to the substrate.

By cutting off the first layer of core along with the veneer, you'll have bending stock with a flexible backing attached. The grain on the first layer of core runs perpendicular to the face veneer, making it very flexible.

Cutting plywood veneer

Use a 6-inch wide strip of veneer plywood. Make sure your band-saw blade is sharp. Set the fence so that the cut is right at the glue line of the first layer of substrate. Set up a featherboard to keep the piece firmly against the fence.

Run a belt sander with an 80-grit belt across the back of the strip to even out the band-saw cut. Now test the flexibility of the veneer. It should bend easily. This process can just as easily be used to cut veneer out of solid lumber.

MAKING A FEATHERBOARD

A featherboard is nothing more than a wooden spring. Many types of featherboards are commercially available, but it's fun to build your own. It makes little difference what angle or thickness you use for the featherboard. The idea is to make the feathers thin enough to be springy but thick enough so they won't break. I have several featherboards around the shop, but here is my favorite.

Start with any piece of 3/4-inch solid hardwood measuring 4 to 5 inches wide by about 12 inches long. Cut an angle of about 45 degrees on one end. Measure down the board about 3 inches and draw a parallel line. Mark out lengthwise lines about 1/8 inch apart. Band-saw down the lengthwise lines to form the feathers.

You can clamp the featherboard in place, but I like the idea of using the table-saw miter slot. Mine is a T-slot. I cut a piece of flat metal to fit into the slot and the tapped holes in it—this allows me to bolt the featherboard anywhere I like on the saw. If you prefer to clamp, simply make the featherboard longer so you can reach it with the clamps. ■

Cutting the feathers on the band saw isn't hard as long as you remember the drift angle of the saw (pg. 59). Careful layout helps keep the feathers the same size. Each should be the same length as well, so angle where the cuts stop parallel with the edge of the featherboard.

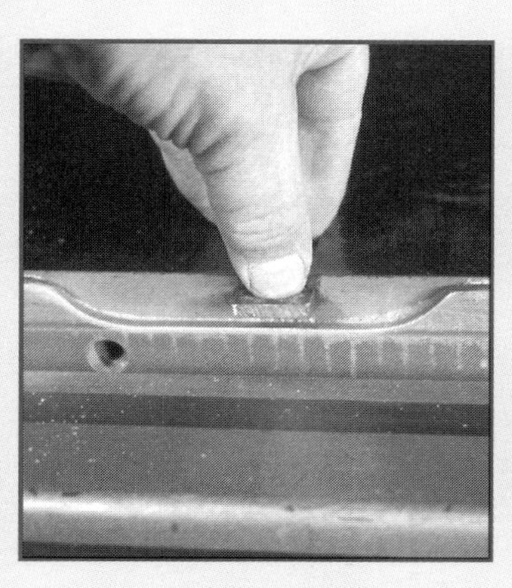

If you're fortunate enough to have a T-slot in your table saw, you're in luck. You can fit a metal slide into the slot for your featherboard.

Drill and tap holes in the metal slide. Then insert threaded rod flush with the underside. Use a center punch to fasten the rods in place with indentations.

The slide with bolts attached is a great way to fasten a featherboard wherever you want one.

Press the featherboard into the workpiece hard enough to bend the feathers a little, but not so hard that it's difficult to feed the board. Hold it in place while you snug the nuts down.

PLYWOOD TUBE SUBSTRATE

Plywood tubes and other curved shapes are commercially available in many sizes. Once veneered, they look like solid-wood cylinders. The process works best with paper-backed veneer. The veneer can be glued to the cylinder as long as the cylinder measures at least 6 inches in diameter and the grain runs lengthwise. Paper-backed veneer is convenient to use, since the veneer edges are already glued together—the backing material keeps the veneer stable and it bends uniformly.

Draw a line down the length of the cylinder. Cut the veneer so it's oversize both in length and width by about 1 inch in both directions. Apply contact cement to both the veneer and the cylinder following the instructions on the container. Carefully align the edge of the veneer with the line. Being extremely careful to avoid any air pockets, roll the cylinder to attach the veneer. The edges should overlap by about 1/2 inch.

With a straightedge and a sharp utility knife, cut through both layers of veneer. Pull back the outer layer (the one on top), and carefully remove the strip of veneer below. Apply yellow glue to the seams. Press the seams together with a small wallpaper roller, making sure the veneer is securely adhered. Carefully sand off any remaining glue, then sand the entire cylinder lightly by hand with fine sandpaper.

A word of caution if using water-based finishes with this technique: The water in the finish may expand the veneer to the point where the veneer wrinkles. If you plan to use a water-base finish, attach the veneer with plastic resin glue. ■

One way to make a cylinder is to start with a prefabricated plywood tube. These are available through mail-order catalogs. (See Appendix.) Use contact cement to hold the paper-backed veneer in place.

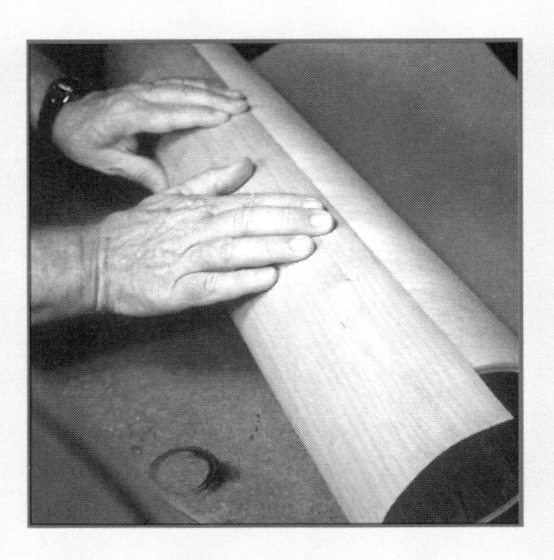

Once wrapped with veneer, the cylinder will look like solid hardwood. Apply the cement and simply roll the veneer onto the cylinder. Overlap the edges where there is to be a seam.

Cut through both layers of veneer with a sharp knife, using a straightedge to guide the cut. If you cut carefully, the two veneer edges will mate perfectly.

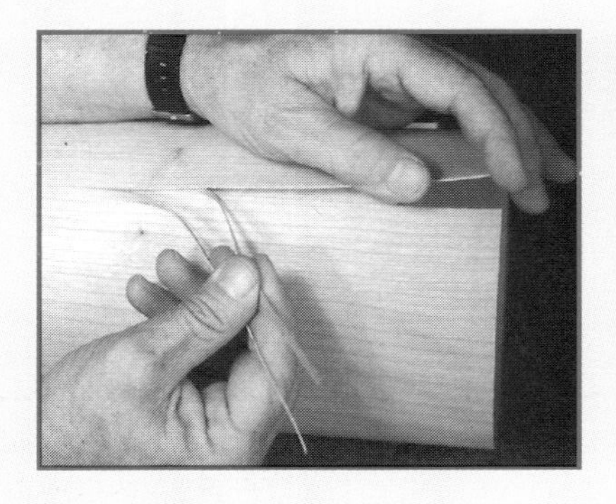

Remove both cutoffs. You'll have to open the seam to remove the one below—peel it up carefully.

The end result is a tight seam where the edges come together. I apply a bit of yellow glue to the edges before I press them down in place for the final time. A light sanding and the cylinder is ready to be trimmed to length.

Milled substrate

A substrate can be built in any number of ways, not just by kerf-bending. For example, a solid block can be quickly and easily stacked and glued from pieces of plywood or MDF, then curved to the desired shape. As is the case with a kerf-bent substrate, the veneer will completely hide the inner structure.

Cut just about any shape you like out of a solid panel. See the section on milling a fair curve, pg. 84. Using the pattern-cutting methods described, you can rapidly turn out a load of identical pieces, which you then stack and glue with ordinary woodworker's glue. Once the glue cures, sand the shape smooth and veneer it.

I use a lot of narrow pieces to make a coopered panel, with the result that there is little shaping to do once the glue dries.

COOPERED PANELS

Coopering is the process whereby narrow boards, called staves, of solid wood are milled, glued up, and smoothed into a curved shape. While you might associate the coopering technique with wooden buckets and butter churns, adding a coopered panel to a woodworking project can quickly raise it above the ordinary. A curved top on a simple chest, for example, can add a great deal of visual interest to a piece. A cabinet with curved doors suddenly comes alive.

When designing a coopered project, you should strive to keep the pieces that make up the coopered panel fairly narrow. Remember, when staves are wide, lots of material has to be removed to round the wood. Not only is the process of shaping wide staves labor-intensive, the wood loses thickness as you shape it round.

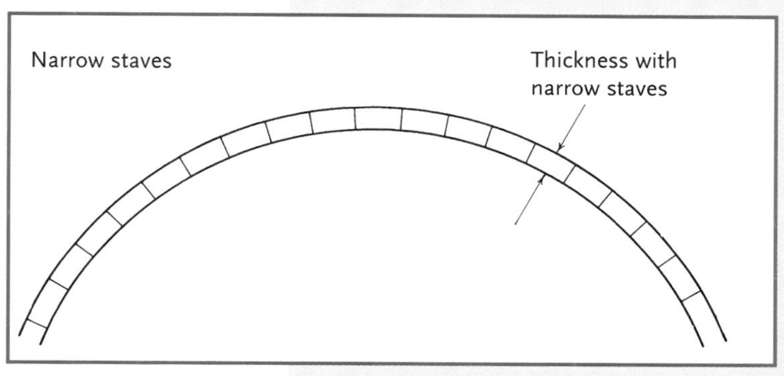

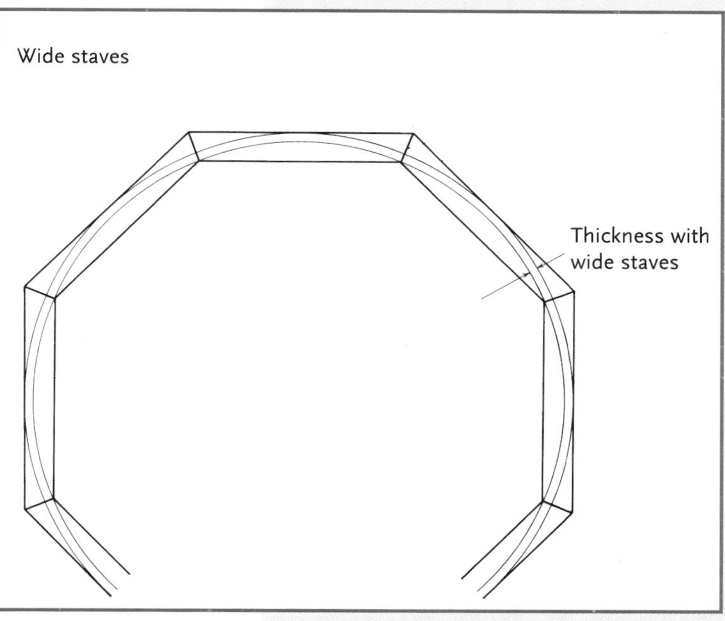

Narrow staves allow greater thickness once the inside and outside faces are rounded than do wide staves, given the same thickness and the same radius.

The essential first step in coopered construction is to make a simple, full-sized drawing of at least the end view showing the radius in rough and finished form. This should take no more than a few minutes of

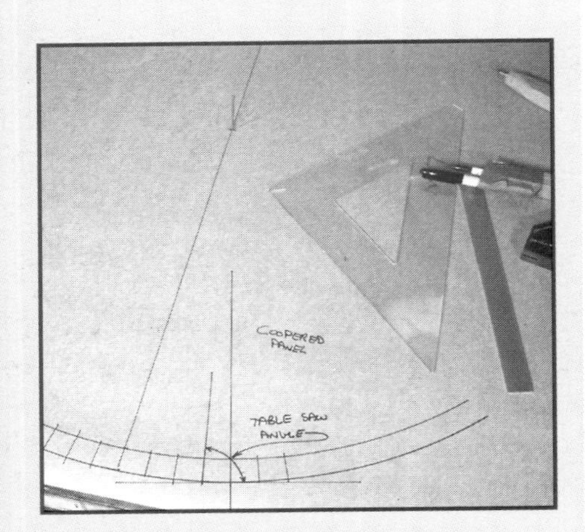

The first step in making a coopered panel is to draw it out full scale.

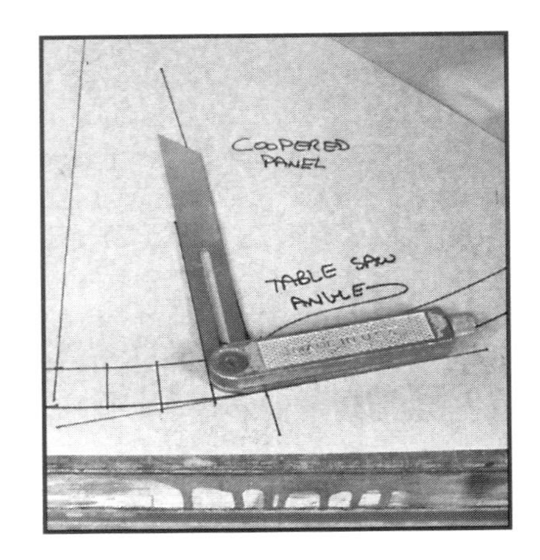

From the drawing, it's a simple matter to measure the bevel angle with a sliding T-bevel.

drawing on paper or an extra piece of plywood, yet it is a critical step that will guide the entire project.

First, draw the inside radius—in the example I'll use for this book, 18 inches. Then determine the width of the panel, in this case, about 12 inches. This information will, of course, depend on your particular project, but the principles will be the same. Draw a line down the center, from the center point to the arc. Now draw in the thickness of the material, which is 3/4 inch in this example.

Now determine the width of the pieces. As I mentioned earlier,

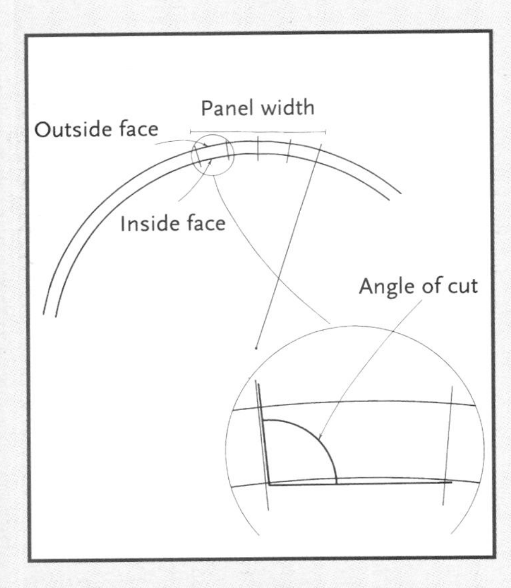

Each stave of the coopered panel has a very slight angle cut on each of the edges. To find this angle, simply measure it right off the drawing. The magnified detail in the circle shows how to measure. Keep in mind this only gets you close. By measuring with a sliding T-bevel, you can set the table-saw blade to the rough angle. Fine-tuning with a bit of trial and error will get the angle spot on.

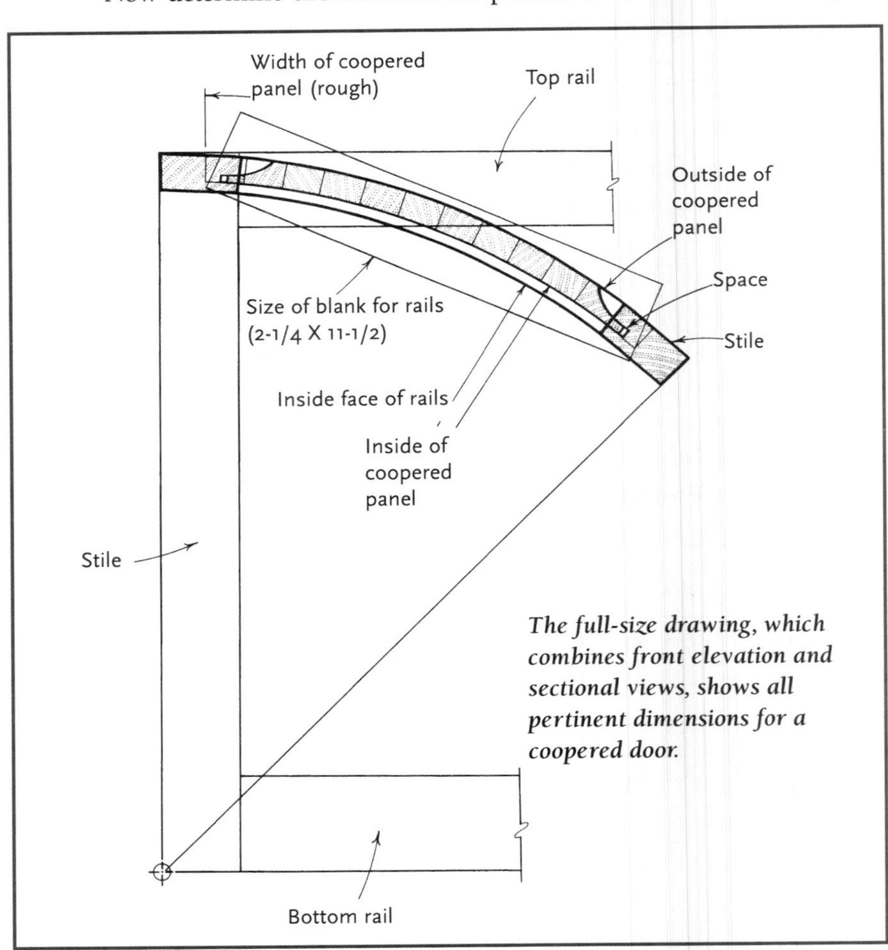

The full-size drawing, which combines front elevation and sectional views, shows all pertinent dimensions for a coopered door.

when I glue up a coopered panel, I like to use relatively narrow staves. Narrow staves can be cut at an angle that achieves the correct curvature more naturally than wider boards would. This reduces smoothing work when finishing up the project. Most important, though, is the fact that it is possible to retain nearly the entire original thickness by using narrow staves.

For our example I'll make the inside face of the stave 1 inch wide. Use dividers to lay out the staves, starting from the center and working outward. The more accurate the layout, the more useful the drawing will be—make sure the drawing allows extra width and length on the material to trim away later. Using a straightedge with one end on the center point and the other on the divider pricks, draw the angles for each stave.

The drawing at left shows that just over 12 staves will be needed to make our sample panel 12 inches wide with an 18-inch inside radius. (Always glue up at least a couple of spare staves to be on the safe side.) Leave 2 inches extra for crosscutting to length once the panel is glued up.

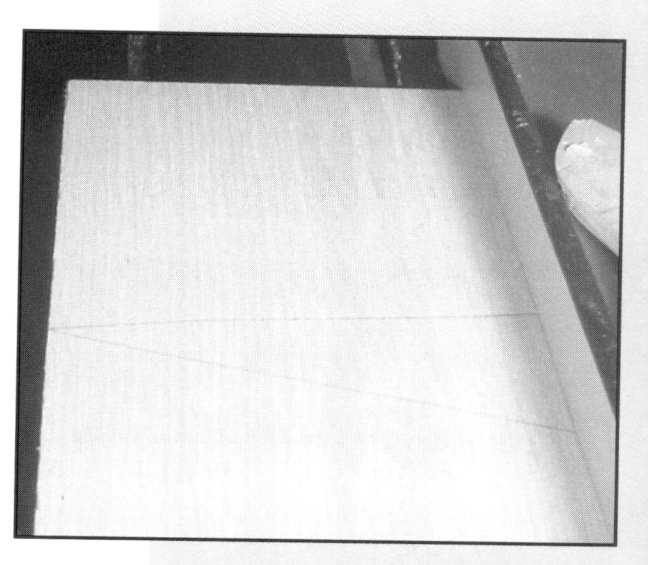

Place a pyramid mark on the board before you start cutting. This will keep the pieces in sequence and right side up.

When the pieces are all cut, the mark you made when you started helps keep them organized.

Cutting staves on the table saw

What follows assumes that you can produce glue-quality cuts (meaning cuts that need no improvement on a jointer), on your table saw. If you can't, invest some time in tuning your saw. With a good blade, a flat table, the blade in alignment with the miter slots, the fence parallel with the miter slot, plenty of power, and smooth feeding, you should be able to make high-quality cuts. If your cuts are of poor quality, you're bound to have a disappointing coopering experience. In this case, take some time to bone up on table-saw tune-up and cutting technique. Practice on scrap until you get glue-quality cuts. It's not that hard to do as long as the saw is working well. Make a couple of comfortable push sticks while you're at it, because you'll be cutting narrow stock.

To make our 12-inch wide panel, you'll need three 6-inch wide boards. Take time to match the grain of the boards, then pencil a "V" pyramid across the faces. This way, after you cut them, you'll be able to reassemble the boards so that the grain looks like it did before cutting. Saw the boards 2 inches to 3 inches longer than needed. In the case of the example, the panel is going to be 24 inches long, so you'd rough-crosscut the boards to 27 inches. Find some scrap wood the same thickness. The purpose of the scrap wood is to adjust the angle of the table saw. Practice on scrap until you get it right.

Finding and setting the exact angle of each stave is best done through trial-and-error. Those who are mathematically inclined may be tempted to figure out the exact angle to three decimal places, but it really isn't necessary, especially since the gauge on the table saw isn't accurate enough to set the angle. It's best to measure the angle right off the drawing with a sliding T-bevel, use the bevel to angle the table-saw blade accordingly, then make final adjustments using trial pieces.

To make the first trial piece, cut a piece of scrap with the fence set about 1/4 inch wider than actual width. After the first cut, keeping the top surface on top, turn the board end for end, and bring in the fence toward the blade to make the second cut at the stave's net width. (You're ripping a narrow piece, so use a push stick to make sure the operation stays under control.)

Crosscut this test stave into pieces about 1 inch long. Place the

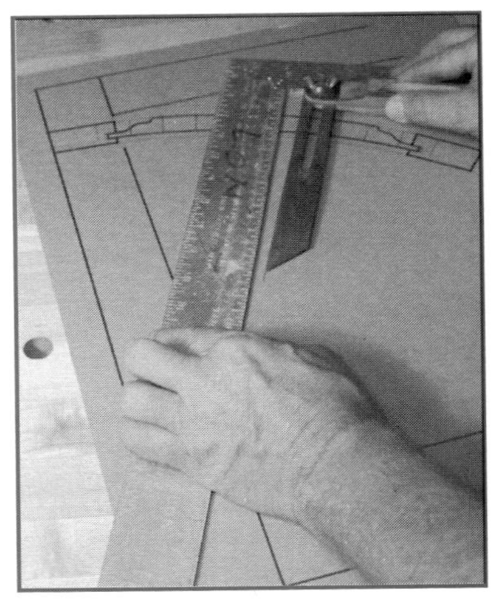

Working on your full-size drawing, take off the stave angle with a sliding T-bevel.

The T-bevel only sets the blade to the ballpark angle for the cuts. Trial and error does the rest.

LEFT-TILTING BLADES

Coopering staves is one of the many processes where a left-tilting table saw comes into its own. With the blade tilting to the left, away from the fence, the angle of each stave can be cut right side up. By "right side up," I mean that the outside face of the curved panel is up, resulting in a cleaner cut on this important side of the board. Reversing the board end for end places the newly cut edge of the board against the fence up off the table.

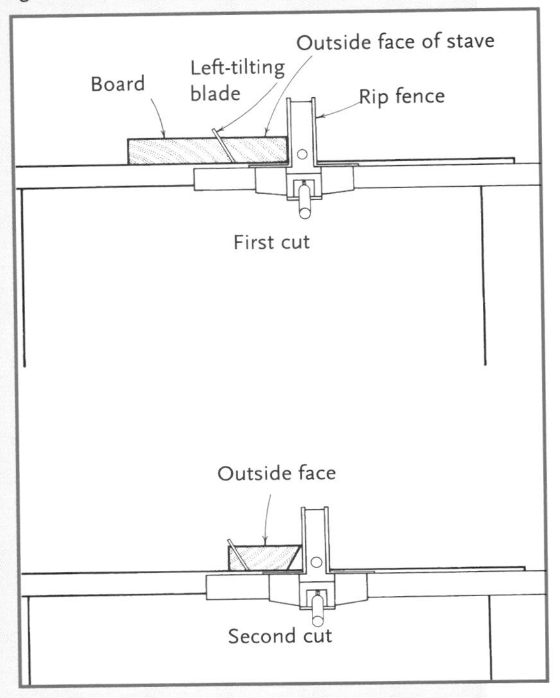

A left-tilting blade lets you saw the staves with the outside face of the stave—the surface that will show—facing upward, which results in a smoother cut.

If your saw has a right-tilting blade, I suggest you move the fence to the left of the blade. This may feel awkward at first, but offers all the advantages of left-tilting saws, including greater accuracy and safety—cutting on an angle with the blade tilted toward the fence traps the piece in such a way that makes the cut far more dangerous. In addition, the piece would have to be cut upside down. When cutting the second side, the sharp angle tends to slide below the fence and get caught, making the cut both hazardous and inaccurate. ■

pieces on the bench so that the top surface is up and the angled cuts fit tightly together. Apply two or three layers of masking tape across the tops of the pieces. In effect, you've made a small tambour door.

Carefully lift the taped pieces off the bench and gently pinch the two end pieces in toward the center. Just like magic, the pieces assume their natural curvature. Holding the pieces together, place the assembly over the drawing. If you're very lucky indeed, the natural curvature will match the drawing. If you're like most woodworkers, the natural curve will be either too cupped or too shallow.

Too much cupping means the angle on your saw is too steep. Move the blade ever so slightly back toward vertical. If the sample is too shallow, move the blade away from vertical to a slightly steeper angle. Repeat this procedure until you get a sample that matches your drawing. Then lock the angle adjustment on the saw. That's all there is to getting the angle just right. I'd be surprised if it took more than two or three tries to get the angle perfect.

To make the table-saw cuts, you need to start out with one straight edge. Move the fence an extra 1/4 inch or so away from the blade so that the first cuts are oversized by that amount. Using a push stick, rip all the boards on the locked angle at this width. Keep the boards in sequence with the top surface up. Now rotate the boards end for end and move the fence in toward the blade to the stave's correct width, and cut all the angles on the other edge of all the pieces.

Your end result should be 14 staves cut at the perfect angle with edges ready for gluing. If you're curious, you can try taping the staves like you did with the samples just to check. I'd bet they're nearly perfect.

Cut a single stave of scrap, then cut it into short pieces. Run some tape across the pieces to keep them together.

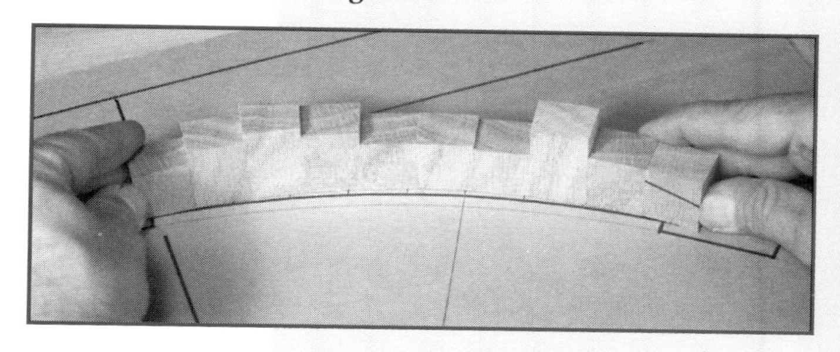

When I place the pieces on the drawing they form a curve, but the radius is tighter than I want. Is the blade angled too steeply or not steeply enough? You guessed it. The blade has to go slightly back toward vertical. How much? Move it a smidgen and try the process again.

Eventually, through trial and error, you'll find just the right angle on the table saw. The natural curvature the pieces take when pressed together by hand will match the one you want on your drawing. Now you're ready to cut the real pieces.

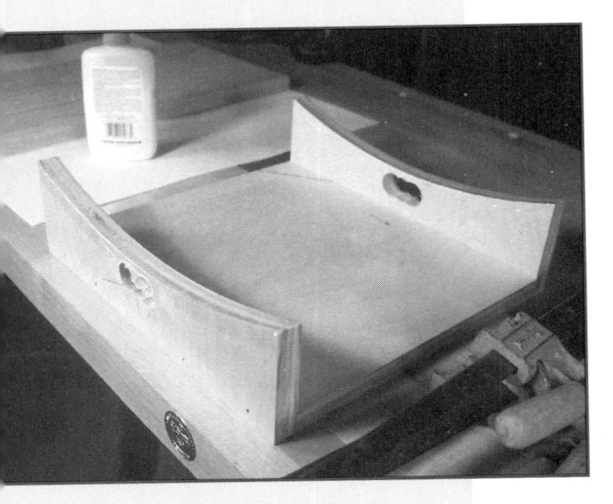

Gluing a coopered form isn't difficult as long as you use a female jig like this one. Just refer to your drawing to find the radius, cut both pieces at once on the band saw, and nail the jig together. Apply cellophane tape to the curved surfaces so the pieces won't stick to the jig.

Making a gluing form

To keep a coopered panel under control, you must glue it up on a form. The form I use is simple to make—it's just a panel that conforms to the drawing—and guarantees predictable results.

Consult your drawing of the project. By drawing a rectangle around the outside of the circle, you can get the shape of the form you'll need to glue up the panel. Draw in lines perpendicular to the end staves. This is the direction we're looking for with clamp pressure. If the panel were flat, the clamp pressure would be directly on the edge, in line with the panel. Since the coopered panel is curved, it means the clamp pressure must be in line with the outside staves. Applying pressure in the right direction forces the assembly down into the mold. Sketch in some pockets for the clamps, designing the pockets around whatever clamps you have handy. Drilling holes in the jig for the clamps is fine if the clamps you're using are too short to span the jig.

This situation calls for a female mold. Cut the radius as carefully as you can on a scrap of plywood. To make the mold more accurate, you can nail the mold pieces together and cut them all at once on the band saw. Cut right on the line as carefully as you can. A sharp band-saw blade and steady pressure will make the job easier.

Nail a female mold piece every 12 or 18 inches along the plywood—the more pieces the better, but no less than every 12 inches. Make sure to use a full bottom piece of plywood, which will keep the jig square and solid. Mark the center line on the mold pieces. Measure out from either side of center and mark the outer limit of the panel. It's important to keep the panel in line with the center. Otherwise, you'll have a panel that is curved just right but has a twist that will drive you nuts.

Full bottom piece of plywood
Female form pieces
Every 12"-18"
Clamp pressure
Clamp pressure
Plywood
Clamp

The female mold should be as long as the panel and as wide as the total of all the staves plus a bit extra.

The last thing to do is to cover the radius pieces on the jig with plastic packing tape. This will keep the panel from sticking to the jig but, more importantly, the tape will allow the pieces to slide freely. Remember, once clamping begins, the pieces need to squeeze together, which means they need to be able to slide.

A trial clamp-up

I recommend that you clamp up all of the pieces at one time on the jig before you spread glue on the pieces. This is not as daunting as it might seem as long as you have a plan.

On the bench, position all the pieces face up and in the sequence ordained by the "V" mark. Then place the pieces in the jig face down. Mark the concave or underside with a "V" drawn with chalk (not pencil—you want to be able to tell the two marks apart). When all the pieces are face down in the jig, check that you can see the alignment marks on the jig.

Now place all the pieces on edge as though you're about to spread glue. Remember that each of the outside pieces will get glue on one edge only. All the other pieces will get glue on both edges. Pretend to spread glue. Turn all of the interior pieces over and pretend to spread glue again. Then place the pieces face down and in sequence. See why you chalked a "V" on the bottom of the panel? It allows you to put the staves in sequence quickly, making sure they're right side up and not swapped end for end.

The next step can be tricky if you work alone like I usually do. The clamps oppose one another and push in opposite directions; the first clamp must stay put until the second takes up the slack. With a little practice (or another pair of hands), you'll be able to control the process. Clamp the whole panel until all the clamps are snug. Then tighten one and loosen its opposite until you have the panel centered and clamped.

This is a good time to assess the joints. With the edges freshly cut there is no need for splines or biscuits. The jig takes care of alignment and the glue takes care of strength. The pieces should fit together tightly, with no gaps anywhere along the panel. If there is a gap anywhere, you might try applying an additional clamp across the two edges, but if that doesn't work, you may have to re-mill the two parts. If you do indeed discover two pieces that don't fit well, congratulations! This is precisely why you clamp up the panel dry.

Now that you're certain that the pieces fit, you know you have enough clamps, and you have a little practice handling the jig and the process, it's time to spread glue.

Notice how the two opposing clamps work in tandem, applying pressure in just the right direction. Not only are the pieces being clamped with lots of pressure, they stay in just the right curvature.

Glue-up

Make sure the temperature of your work area is at least 60 degrees; any colder and the glue may not cure properly. If you need to, it's easy enough to make a little tent with a light bulb inside to keep the glue, the jig, and the clamps warm (pg. 122). You can use either white or yellow woodworker's glue. White glue sets up a little more slowly than yellow glue, allowing a little extra working time, but with either glue you still have to move fast.

Place the pieces on edge and spread glue on all the edges. Rotate the end pieces so that their interior edges are up. Now turn all the interior pieces over and spread glue on their other edges. Quickly place the pieces in sequence, aligning the V-shaped chalk mark. For optimal strength, the pieces must be in contact before the glue has a chance to skin over.

Apply the clamps snugly but don't tighten them down. Align the panel with the centering marks as you go. Gradually tighten the clamps until they're about as tight as you can achieve using three fingers. Even with small clamps, this is as much pressure as you'll need. When you're finished, mark the time on the piece with chalk, so you can keep track of when you should unclamp the assembly. Take a break and admire your work.

The label of the glue I use says to clamp the pieces for 10 to 30 minutes. To be on the safe side, I usually leave the clamps on for an hour or so, depending on the temperature, but no more. If the glue you use suggests different clamping times, by all means follow the directions.

After about an hour, the glue squeeze-out will be soft and fairly easy to clean up. Although it might be tempting, don't try to remove the excess glue with a wet rag. This will just smear diluted glue all over the surface and into the pores of the wood. Although the surface may look clean, any glue residue will become painfully visible during finishing, when it interferes with the action of your finishing material. It's far better to scrape off the glue with a hook scraper.

Carefully remove the panel from the jig—remember, the glue has set up but it hasn't yet cured, so it's entirely possible to break apart a glue joint with, say, an inopportune drop to the floor. It's important to distinguish between set-up time and curing time. The glue sets up in about an hour at the correct temperature. This is also the clamping time. After an hour, the panel comes out of the clamps and excess glue gets scraped off. The glue has "set" but not yet "cured." It's too soon to machine the panel. The glue should cure for at least 24 hours before you machine the panel. It takes this long for the glue to reach its full strength.

When you're ready, cobble up a setup that allows you to scrape both sides of the panel at once, then slowly and carefully remove the excess glue. Some wood will come off with the glue, but this is okay, as long as you don't take off huge amounts. When you've finished cleaning the panel, clean the jig and set the panel back on the jig to cure. Keep the temperature constant.

THE HOOK SCRAPER

There are two kinds of scrapers: hook scrapers and cabinet scrapers. I use both freely—the hook to rough-out work and the cabinet for fine finishing contours. The hook scraper is more commonly known as a paint scraper. But it's actually a very useful woodworking tool if properly tuned. I like Red Devil single-edge scraper blades, for which I make wood handles. Using a fine mill file, I sharpen the edge with a chisel-like bevel until it's razor sharp. In the case of a curve like our coopered panel, it's a snap to shape the blade to match the radius of the panel's concave face. Then not only will the hook scraper take care of the excess glue, it will rough-in the curvature.

Although a hook scraper is the ideal tool for removing soft glue, it works at peak performance only when it's sharp. Since the metal is relatively soft, it dulls easily, which means you need to sharpen the blade more or less constantly. But sharpening takes about 10 seconds. Holding the scraper on the edge of the bench, simply filing downward for three or four passes with the fine mill file will renew the edge to razor sharpness. With the blade this sharp, excess glue will actually lift off the surface, along with a wood shaving that rivals those cut with expensive hand planes.

It's convenient to keep several scrapers on hand. File a curvature on one blade for working on the inside of a panel. Keep one scraper flat or almost flat—round the corners to keep from gouging the workpiece. ■

A hook scraper uses a beveled cutting edge, like a chisel. To file a hook scraper, keep the file parallel to the bench edge. This way you can hold the scraper at any angle you wish.

Use a fine metalworking mill file to make a matching curved shape, which speeds the smoothing of a coopered panel.

The shape and quality of the shavings produced is the true test of a cutting tool. The shaving in the foreground, cut with a tool that cost 50 cents, is as good as you'd get with a more expensive tool.

A sanding sled is useful when you want to sand curves. Sand the outside layer of a piece of 1/8-inch plywood, glue on a section of a new 120-grit sanding belt and a couple of handles, and you'll have a nice, flexible sanding sled.

Smoothing the inside of the panel

The inside surface of the panel is the most challenging to shape, so I like to get it out of the way first. Shape a cabinet scraper as discussed on the following page and use it to smooth the inside of the piece after you've scraped it with the hook scraper.

If you want to sand the surface as well as scrape it, I suggest making a sanding sled. Take a piece of 1/8-inch plywood that's 3 inches wide and about 12 inches long and belt-sand one face until the outer layer of veneer is removed. This makes the plywood quite flexible. Cut a new 3-inch sanding belt to the same length. With contact cement, glue the sandpaper to the plywood. Glue two small blocks to the outside of the plywood for handholds and you have a useful tool for sanding curves. I use 120-grit belts mostly, but having different sleds with different grits is well worth the time to make them. Whatever you do, resist the temptation to hold plain sandpaper in your hand. While you may be able to get the surface so that it feels smooth, lumps and hollows will usually be left behind.

Smoothing the outside of the panel

This face of the panel is easier to shape but is also easier to mess up. Remember, mistakes here will be visible in the finished product. Those 12 coats of varnish you're planning to apply will magnify any dip or bump. The idea here is to keep the entire panel fair, with the curve even and smooth. For this work I typically use a small block plane, the straight side of a cabinet scraper, some chalk, and a thin batten about 1 inch wide. I suggest you resist the temptation to make a scraper for smoothing the outside of the panel. Inevitably, the outside corner of the scraper will dig into the surface no matter how careful you are. If this happens, you'll find that you've just ruined your panel.

If you get a lump or a hollow, it can be easily isolated by using chalk on a thin batten. It will highlight the high spots so they can be removed.

Mark the ridges where the staves come together. Apply chalk to the batten. Bend the batten over the panel, rubbing chalk on all the ridges from one end of the panel to the other. This highlights the high spots.

The cabinet scraper, ground to the inside radius, does a great job of making the panel's inside contour smooth.

SHAPING AND SHARPENING A CURVED CABINET SCRAPER

For novice woodworkers, one of the most baffling and frustrating parts of woodworking with hand tools is getting a cabinet scraper to work properly. But like most things that seem complicated, break the process down to the basics and it becomes fairly straightforward.

A scraper cuts because the burr on its edge acts as a microscopic chisel to remove fine shavings from the wood. In this case, the edge of the scraper is not only burred but curved. Using an inexpensive stock cabinet scraper, tape a scratch awl to a compass and mark a single line along the curvature you need.

At the bench grinder (or with the same mill file you use to sharpen your hook scraper), carefully shape the scraper. Have a pail of water handy to keep the metal cool. Grind a bit, dip the scraper, and grind again. Don't allow the metal to become hot to the touch.

Once you've ground the scraper to the proper shape, run your fingers along the curve and make it as smooth as possible. With the scraper in a vise, slowly draw the mill file along the edge, smoothing and removing the grinding marks. Ease the corners slightly so they won't dig in. Work slowly and carefully until the edge is square and smooth.

To hone the edge, I like to use wet/dry sandpaper on glass (a spritz of water helps the sandpaper stick to the glass, keeping the paper flat), but a diamond stone is good, too. I am reluctant to use my sharpening stones for this because I don't want to wear grooves in the stones.

Starting with 220 grit, hone the edge square. If you spray some water on the sandpaper, it will cut more easily. Continue through several grits, from 400, 600, 800, and 1000 up to about 1500. You'll notice that the edge of the scraper is shiny.

The honing of the edge will raise a sharpening burr. Lay the scraper flat on the sandpaper and remove the burr. At this stage, the corners should be square and sharp. Test them by seeing if the scraper will cut. If it slides across the surface of a board without cutting, it's not yet properly honed. Once you can make a cut with each square edge, burnishing is a snap.

The point of burnishing is to raise a burr. Use a burnishing tool lubricated with some light oil—just enough to make the tool slide along easily. It doesn't take much pressure with the burnisher to raise a burr. Three or four passes should be plenty to raise a very aggressive burr.

You can either push the scraper away from you to cut, or pull it toward you. As long as the work is clamped securely, it's a matter of personal preference how you use this tool. I prefer to push, since my thumbs can bend the scraper just enough to get a good cut. When I pull, I can't bend with my fingers nearly as effectively. It takes some practice, but once you get the hang of it you'll find that scrapers are marvelous tools, especially for smoothing the interior of a coopered panel. ■

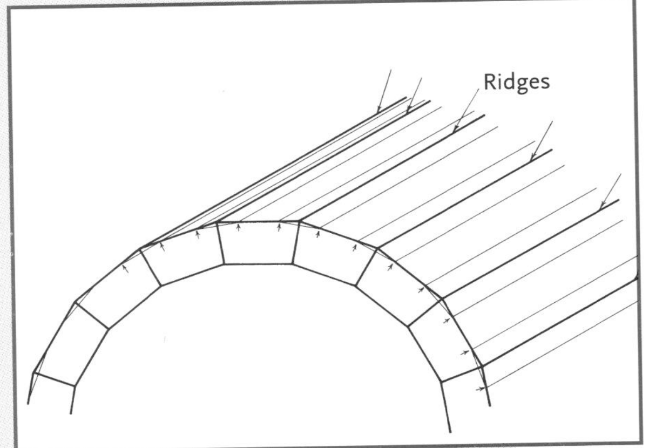

At this point you have a curved panel consisting of several flat pieces. At every joint, there is a slight ridge. The goal is, after planing with a block plane, to have all the ridges small and the same distance apart.

With a block plane tuned to shave thin, tiny shavings, make a pass or two along the marked ridges. Keep the plane balanced on the ridges. As you work on the ridges, notice that you're removing the chalk marks.

As you float the plane along the ridges, try to divide the angle between the two adjacent pieces. When you have each of the ridges planed down along the entire width of the panel, the result should be equally spaced ridges.

Once again chalk the batten, mark across the width of the piece, and repeat the process, again dividing the angle between the two adjacent pieces. Before you know it, the panel will be perfectly rounded with no dips or lumps, just small ridges.

Once the ridges become so small that they're difficult to control with the plane, use the flat side of the scraper. Go as far as you are able with this tool, then use the sanding sled (pg. 184) to remove the last of the ridges and chalk marks.

This curved raised-panel door is coopered from solid-wood staves; the curved rails are milled from solid stock, as are the straight stiles.

MAKING A CURVED RAISED-PANEL CABINET DOOR

Making a curved cabinet door isn't difficult if you follow some easy steps. But it isn't a project for a beginner. You should be familiar with raising flat panels before you attempt a curved one. I'll demonstrate by building a convex door, but one that is concave would involve the same techniques. The panel itself will be coopered, that is, glued up from solid-wood staves with beveled edges. The vertical members of the frame, or stiles, will be straight. The horizontal parts of the frame, or rails, will be curved. We'll mill both rails and stiles out of solid wood. Notice I didn't mention steam-bending nor bent lamination. Neither technique is necessary for making this door.

You could make the rails by steam-bending or bent lamination, but whenever possible I would choose milling.

It's fast, accurate, and gets the job done right the first time with no learning curve. The door we're going to make spans a 45-degree arc. Two of them span a full 90 degrees.

If the panel was flat and not raised, you could make it by bent lamination. But raising it means we'll be milling a profile around the edges, which would expose layers of laminates in an unsightly way. The coopered panel will look like a solid board that's curved and like milling the rails, the process gives us an accurate radius. To build the panel itself, refer to the previous section.

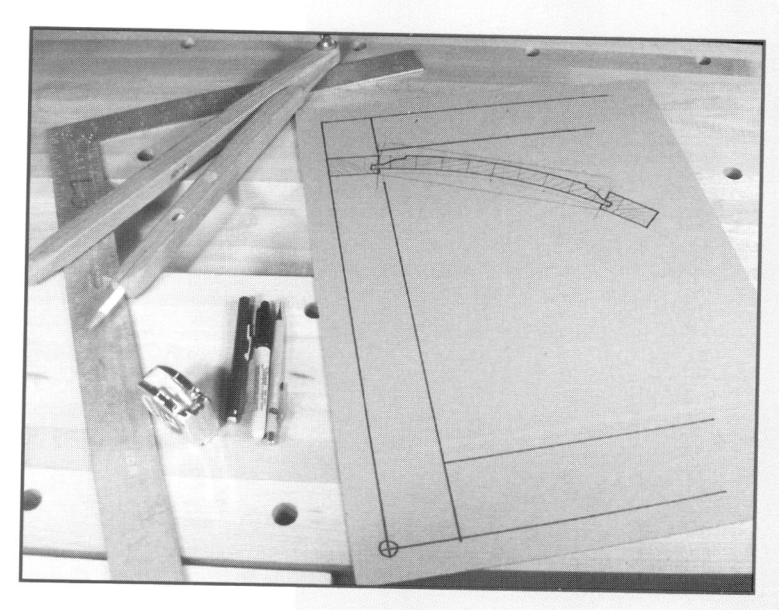

This is all the drawing you need to make for laying out the curved door. This drawing shows full-size the plan view of the curved rails, the panel, and the straight stiles, plus the height of the door showing the top and bottom rails. To figure out how big of a board you need, simply draw a rectangle on top of the plan view.

Drawing and layout

A full-size drawing is a must. You really only need to draw the top view looking down at the curvature of the door, along with a front view of one stile showing the heights of the panel and rails. Draw the stiles, the rails, and the curved panel showing how the panel fits into the slots in the stiles. Draw a rectangle around the rails. This shows how long and wide the board must be for the rails.

On the same drawing you can figure the widths and the angle for the staves that will make up the coopered center panel. You can also sketch in the details of the joint between the rails and stiles.

Don't forget to leave room for the panel to expand and contract. That means the tongue must clear the bottom of the slot by 1/8 inch or so on either side. The panel must have room to float inside the frame or it will blow the frame apart. I usually make the slots in the stiles and rails 1/4 inch wide and 3/8 inch deep.

The panel

Once you get it glued up, it's straightforward to trim the panel to size using a regular crosscut sled on the table saw. Make sure you allow enough width for the tongues that will fit into the stiles and rails, and remember, there's a tongue on both sides.

The stiles and rails

We're going to build a simple, Craftsman-style door that doesn't have a profile milled onto the edges of the rails and stiles. This simplifies the process and the discussion. Once you understand how to proceed, you'll find it easy enough to use a cope-and-stick system. This is where a male profile milled onto the edge of the stiles fits into a female profile milled on the ends of the rails. The door we'll build has stiles and rails both 1 inch thick, with a panel that is 3/4 inch thick.

The stiles we can simply mill out of a straight board. The rails are a bit trickier. The bottom rail is 2-1/2 inches wide and the top rail is 2 inches wide. To simplify the process, I'll glue up a single block 2-1/2 inches thick, then band-saw both parts out of it.

I usually lay out and band-saw freehand if I have only a couple of parts to make. If you like, refer to Part 2 for information on milling circular parts; it's a more accurate way to mill the rails. After band-sawing, I set up my poor-man's edge sander, just a belt sander on its side, to smooth off the saw marks and fine-tune the shape of the rails. (See Part 2 for details on fairing a curve.) Rip the top rail down to 2 inches in width. Be sure to save the cut-off pieces for later setups.

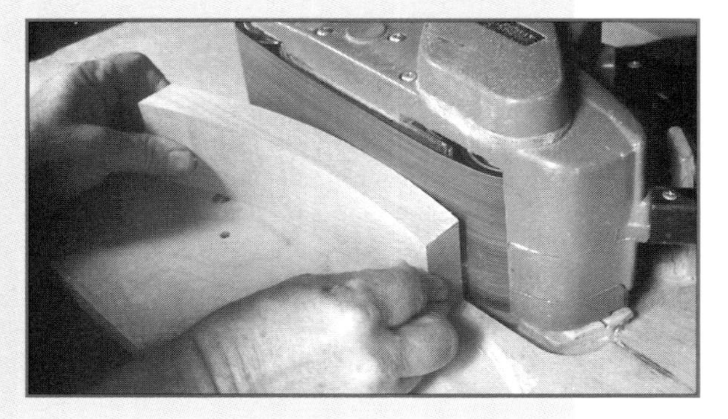

When I have only a couple of parts to make, I lay out and band-saw the stiles freehand and belt-sand them smooth.

By using a cut-off scrap from the rail blank, it's easy to cut the ends at the correct angle. Position the workpiece carefully, and use the curved scrap to hold and clamp it in place on the bed of the chop saw. Check it on the drawing to be sure you are sawing the correct angle.

Milling slots in the stiles and rails

To cut the slots in the curved rails, use a piece of the cut-off as a guide and set up a 1/4-inch diameter straight bit in the router table.

It's time to trim the stiles and rails to length and run the slots for the panel. To hold the curved rails accurately, I use a curved scrap of the material from which they were sawn. By positioning the scrap in the chop saw, it's easy to cut the ends of the rails to the correct angle.

I usually cut the slots in the straight stiles with a dado setup on the table saw, but the curved rails are a different story. The slots on the curved rails must be cut with a router. Once we have all the slots cut, we can size the tongue on the panel to fit.

Set up your router table using a 1/4-inch diameter straight bit. Position the cut-off guide to allow for 3/4 inch between the outside cutting edge of the cutter and the guide. The idea is to position the slot so that the face of the curved panel comes out flush with the stiles and rails. I like to cut the slot about 3/8 inch deep. Take several passes, raising the bit a little each time until you reach the full depth of the slot.

The slots in the stiles can be cut on the router table or the table saw. In this case, since we're using the router already, we'll cut the stile slots there also. Use the same setup minus the curved guide. Make sure the slots in the rails line up with the slots in the stiles.

I like to reinforce the joint with screws rather than dowels. The screws make sure there's no way the joint will ever open up even after heavy use. Counterbore the holes, cut matching plugs, and sand smooth.

Making and raising the panel

Raising a curved panel is not a lot more hazardous than raising a flat panel, but hand-feeding a raised panel whether flat or curved requires getting a feel for the job. Rehearse the setup with the machine turned off, and please make sure you use all the guarding I suggest to minimize the risk of injury should something go awry. Always make several light passes

Raising a curved panel isn't a lot different from raising a flat one, but given the setup, it's not for beginners. This setup uses the scrap from bandsawing the rails to make a cradle.

instead of one or two heavier passes. Having said all of the above, cutting the curved panel isn't all that different from cutting a flat one.

As you can see in the photo at left, the panel slides along its curved face on the curved cradle. The end of the panel rides along the fence. By raising the bit a little at a time, you cut deeper and deeper until the tongue is the right thickness to fit into the slots in the rails.

Shaper vs. router table

I raise panels on the router table mostly because I think it is safer. That's because shaper cutters are designed to run above the panel, trapping the workpiece between the cutter and the table surface. The shaper cutter could be lowered down into the cut gradually, but if the panel were to hit a bump or chip and rise just slightly off the jig, a deeper and perhaps disastrous cut would result. Router bits, on the other hand, are designed to cut from below. This makes it possible to cut gradually into the panel by taking a series of light cuts, raising the bit just a little for each pass. Any bumps or chips will lift the panel off the cutter rather than driving it deeper.

The shaper can be set up to work the same way as the router table, but the cutter's rotation must be reversed. The shaper is one of the few woodworking tools that is reversible, making this operation much less hazardous, though you do need to get used to the idea of feeding the work into the cutter from the opposite direction (left to right). If you are going to use the shaper to raise a panel, I think reversing the cutter and running it below the workpiece is inherently safer than running it from above.

The shaper cutter at the top of the photo is designed to run above the panel. The router bit is designed to run below it. It's a lot safer to run the cutter below the workpiece.

Assembling the door

Dry-fit all the parts, making sure everything goes together easily. Remember, the glue will swell the wood slightly so a loose fit is fine. The panel must have room to float, so I put little pieces of rubber inner-tube or cork into the slots to position it before assembly. It's okay to glue the center inch or so of the panel to the rails, but other than that the panel must not be attached to the frame.

Glue up the stiles and rails and gently clamp everything together. Add the screws if you're going to do that. Measure across the corners to make sure the door is square. And there you have it—a curved raised panel door.

Cutting the straight edges of the panel is a lot less hazardous than cutting the curved ends. It's important to "feel" the cut to make sure it is working safely. The board fastened to the router table keeps the panel at the proper angle.

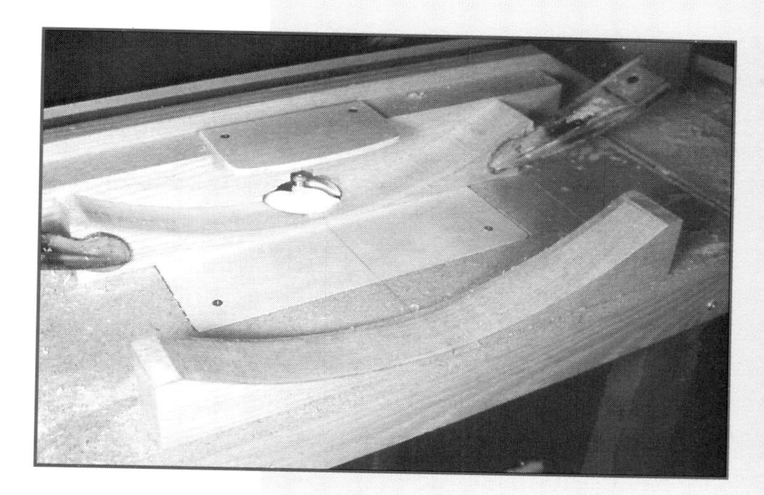

Note the small plywood guard mounted on the fence above the cutter. It isn't a perfect guard, but at least it reminds me of the location of the cutter even when it is hidden by the work. A light spray of lubricant such as Top Coat makes the panel glide over the jig for a smoother cut.

Appendices

A SUMMARY OF MY SHOP SAFETY RULES

I have large arrows on my table to remind me which way the cutter spins.

Avoid distraction; phone off the hook, door locked.

Keep fingers where you can see them and out of the line of cut.

Keep both hands on portable tools.

Small clamps often loosen from vibration and should not be used.

Use high-quality and sharp blades, bits, and cutters.

Recheck bolts and collets for tightness.

Don't use big cutters in small machines.

Move deliberately, stay focused.

If something doesn't feel or sound right, stop working.

Keep tools unplugged unless you're about to turn them on.

Make sure you have good light and footing.

Make off-switches easily accessible, foot-operated if possible.

Loosen the belt when hand-feeding. ■

APPENDIX 1:

Machine Safety

I PUT SAFETY BEFORE CONVENIENCE. I put safety before speed. I put safety before cost. It is not worth it for me to do a job faster or more profitably if it means taking a risk that is within my power to eliminate or at least minimize.

Surely this can go too far in some areas for more experienced hands who are comfortable doing things a certain way, but I feel that any expense or time required to make an operation safer is justified.

This section is intended as a quick guide to remind you of the things you should never forget. You'll find detailed information on safety principles in all the parts of the book, where safety runs as a strong theme throughout the pages. What follows here tends to focus on shapers, but the information can as easily be applied to operating any woodworking machine.

TIME, SPEED, AND YOUR NATURAL PACE

Woodwork demands time. Lots of accident stories start out with a reference to being in a hurry. There is always a time factor with any work, of course, but if you try to work faster than your natural pace, it will almost always result in a mistake that takes more time to fix than you thought you would save by hurrying. And that is precisely when you are at the highest risk of having an accident. One of the hardest parts of woodworking is allowing enough time to do the work safely.

ON LOOSE KNIVES

Most stories about shaper accidents involve thrown knives. One such story is in the sidebar on this page. It is relatively easy to eliminate this common safety issue by using only solid three-wing cutters or router bits in your shaper.

Some acquaintances may think you're crazy to buy a whole new solid cutter every time you need to run a new pattern. To some degree, they'll be right, but if this is what it takes to feel safe, then the expense is well worth it. Three-wing cutters of all kinds are readily available and even when you need a pattern that is not stock, custom-made cutters are not that expensive, especially when the time spent grinding knives is taken into account.

Is it unheard of for carbide to part company with solid cutters? Unfortunately, no. But it is rare enough, certainly more rare than having loose tooling fly out.

WIRING

When I wired my new shop, I deliberately did not hard-wire my machines. When you have your hands in the innards of a machine, especially one that is belt-driven, it's comforting to have that plug and cord draped across the table where you can see it. As a rule, keep your shaper unplugged unless you are about to operate it.

PROTECTION FOR EYES AND EARS

Frequently overlooked are those things that make work more comfortable physically, like good light, a firm and comfortable footing, and eye and ear protection. It is amazing how many professional woodworkers do not protect their eyes and ears. It reminds me of playing football without a helmet.

Eye protection should be obvious, but woodworkers are often reluctant to adopt this basic safety measure until after they have a problem. Ear protection is even less commonly used than eye protection.

A SHAPER STORY

I have spent a good part of my career as a stairbuilder, inventing ways to do things I have never done before. Learning how to run a shaper was not something I had planned to do on short notice. I had contracted to build a staircase thinking that I would sub out the millwork on the handrails. I suddenly discovered, however, that I would have to do the millwork myself if I were going to meet my deadline. I knew that it had to be possible to make things on a shaper safely, but I had heard enough stories to make me a bit nervous at the prospect of operating one. As the following tale illustrates, even experienced hands can have trouble.

On a cold, clear December day in 1985, I met the man who told me the first shaper story I had ever heard. Working alone at his shop, he told me, he was running some oak handrail when suddenly the piece kicked back. Somehow the shock loosened the spinning knives, which started to slowly come out of their mounting. Thinking that at any second the knives would go flying across the shop, he dove behind the table saw. Sparks flew in all directions as the knives ate their way into the rapidly enlarging hole in the shaper's cast-iron table. Fortunately, the knives were partly below the surface of the table so they did not fly completely out, but the racket! He said it sounded like a 747.

But then came the worst part, the knowledge that at any minute the shop could catch on fire from the sparks. Crouched behind the saw, he realized he was going to have to come out from his shelter and crawl back over to the shaper to shut the thing off. He slithered over to the shaper on his belly, reached up, and hit the off-button. Silence was never so welcome.

Finally, as he stood on wobbly legs to survey the damage, he realized what a lucky man he was and how cheaply he had learned his lesson. The shaper had a 6-inch diameter hole in the table where there used to be a 4-inch hole, but other than that no harm was done. A close one indeed. ■

Audiologists (the people who test hearing and fit hearing aids), attribute hearing loss to loud machinery. This loss is so gradual that it often goes unnoticed until it's a problem. Even if you know that you have had some hearing loss, wearing ear protection now will preserve what hearing you have left.

SAFETY MODIFICATIONS

Try using old rubber doormats as footpads for your machines. They work great for firm footing and reduce operator fatigue as well.

You can modify your machines in a variety of ways to make them safer. First, make or buy foot-operated off-switches. You can simply make a paddle on a hinge with a protruding knob mounted in such a way that it comes in contact with the off-button. That way you can keep both hands in position if you have a problem. Your body can remain upright and in balance, and you do not have to look away from the cutting area or use one hand to fumble around trying to find the button and turn off the machine. More and more machines seem to be coming this way directly from the factory.

A foot-operated paddle switch lets you turn off the machine while keeping your eyes focused on the work.

When hand-feeding on the shaper, tension the belt just enough to spin the cutter, but keep it loose enough so that if the cutter should become blocked for any reason, the pulley can spin on the belt. That way the cutter can stop even though the motor continues to run. It takes some trial and error to get the tension right, but it is very easy to adjust.

CHECKLISTS AND NOTES

Pilots use checklists every time they land or take off even though they might have performed the same operation thousands of times before. Why? Because remembering to lower the landing gear (or tighten the spindle nut) is simply too important to leave to memory alone.

Get used to relying on checklists to keep you focused. In a quiet moment, write down the sequence of events for a given operation. That way, the next time you run a piece through the machine you can concentrate on that specific task instead of wondering if you have forgotten something. Include items like tightening the height-adjustment lock and checking the spindle nut along with the more general items like counting the pieces after each operation.

When you quit for the day, simply make a mark on the list where you have stopped. Note things that you discover in the course of the work that may have to be fit into the sequence at some later point. When you return to the work, this will help you regain concentration much more quickly than if you try to work from memory.

This is never truer than for operations you may do only occasionally with a fair amount of time in between. The checklist made while the operation is fresh in your mind is a wonderful resource to have available if a year down the road you find yourself again making the same parts.

Taking photos of operations, especially those involving elaborate setups, is a great habit to get into. Clearly label the parts and keep hardware where it won't be used for something else.

DISTRACTION

Nothing is more dangerous than being distracted when setting up or operating a machine. This is especially true with shapers. When setting up or running a shaper, the shop door should be locked, the answering machine should pick up the phone, and the dog should be exiled to the yard. You may be able to focus just fine in the middle of chaos, but on the off chance that you may forget to tighten some vitally important bolt, you should strive to minimize distraction.

When something comes up, as it often does, to draw your attention away from your work, shut everything off right then. Take a few minutes to write down what the next step was going to be, tape the note to the machine, and go handle the problem. When you return to the machine, take a few extra minutes to refocus.

RESPECT YOUR HUNCHES

Do not hesitate to postpone a tricky or unfamiliar operation if something does not feel right. You have no doubt encountered times when you have checked and rechecked a setup and it looks right but does not feel right for some reason. Usually it is Murphy (as in Murphy's Law) drawing your attention to something you have overlooked.

When you pay attention to these feelings, you are almost always rewarded with a realization that you were about to make a mistake.

ON MAKING LIGHT CUTS

At least at first, take lighter cuts than the machine is capable of. As you warm up to the process and get used to the feel of the operation, gradually work up to heavier cuts.

Lighter cuts take a little extra time but are easier on the machine and the nerves. If an initial pass is not satisfactory, take steps to correct the problem. An initial light cut usually gives you another chance to try the cut without having ruined the piece by going all the way the first time. Besides, once the machine is really humming along, it is so much fun that you usually don't want to get to the bottom of the pile any sooner than you have to.

With some cuts, such as raised panels, the cutter can be progressively raised or lowered into the piece a little more with each cut. Much of this is familiar to anyone who has used a router, but with much more horsepower at work with a shaper, the principle of taking light cuts becomes much more important.

On most machines like shapers and table-mounted routers, the fence is adjustable in and out so that it can be set for progressively deeper cuts. With molding setups you plan to use again, carefully measure the location of the fence for a given cut. In your notebook, make drawings, write down measurements, and even staple photographs to aid in future operations.

PATTERN CUTS

With pattern cuts, I band saw as close to the line as possible to minimize the depth of the actual cut. (Read more about this in Part 2.) The less material you have to remove with the cutter, the smoother and safer the cut will be.

With contour cutters, use bearings with graduated sizes on the same bit so that the larger bearing makes a light cut and progressively smaller bearings make deeper cuts. Once you warm up, you might skip a size on occasion, but this gives you the option of making progressively deeper cuts a little at a time.

SAFETY CUTTERS

Recent developments in cutter/bit design are inherently safer than their predecessors. The European standards for tooling are much more strict than ours; European safety cutters have a closed design limiting the depth of cut. (See Part 2.)

APPENDIX 2:

Glossary

Batten—A batten is a thin piece of wood used to draw or cut a curvature. When using a batten to draw, it should be thin enough to bend but as stiff as possible.

BTU—British Thermal Unit, a measure of heat output of a burner or boiler.

Center point—The point around which a circle is drawn. All points on the circle are the same distance from this point.

Circle—The shape you get with a compass, as opposed to a curve, which is neither a straight line nor a circle.

Circumference—The distance around the perimeter of a circle.

Clamping time—The length of time an assembly must stay under pressure while the glue sets up.

Compression—Compression in this context refers to the wood fibers becoming increasingly compacted by the forces exerted in bending. Unless a board is forced to compress, it is unlikely that it will bend successfully.

Concave—The inner, or cupped, side of a bend or coopered panel. The inside of a ball is concave.

Convex—The outer surface of a panel. The outside of a ball is convex.

Coopering—Thin boards, called staves, are ripped at a precise angle so that when they are glued together they form a curve. The term comes from barrel-making.

Curing time—The length of time glue takes to fully cure, that is, to reach its full bonding strength.

Curve—A shape that is neither straight nor a circle. Any sort of freeform curve falls into this category.

Diameter—This is twice the radius and the distance across the circle through the center point.

Dividers—Dividers look like a compass but have a point on each of the adjustable legs. It is a very accurate and useful tool for dividing spaces equally when drawing.

Elasticity—Elasticity is a property of all wood, but green wood is especially elastic. Elasticity refers to how a piece of wood returns to its original shape if it is bent cold.

Fairing a curve—Fairing is a boatbuilding term that means to smooth a curve so that no lumps or bumps remain. It is done in one of two ways—either by removing the high spots or filling in the low spots.

Flitch—A stack of veneers sliced out of a log and stacked back the way they were cut. Looking at the end grain of a flitch, you can see the growth rings of the tree just as if it were still solid.

French curve—A drafting tool that is curved to make drawing curves easier in small scale.

Grain runout—The grain in a plank is determined by the pattern of the growth rings as it is cut or split. Runout is when the grain of the plank runs diagonally to the edge of the plank.

Kerf—The width of the sawcut made by a blade.

Laminate—Each layer of wood in a lamination-bend is called a laminate.

Lignin—A greatly oversimplified definition of lignin is the pitch, sap, or semi-liquid that remains in wood after it is cured. Lignin is the glue that binds the wood fibers together. When it is heated, the lignin softens slightly, making bends possible.

Lofting—A boatbuilding term that describes drawing a boat full-size on the floor of the loft above the workshop.

Open time—The length of time glued surfaces can be left open before the glue skins over.

Plastic—The property in wood that allows it to bend. When a piece of wood is plastic, it will bend and not return to its original shape once the clamps are removed (except, of course, for a small amount of springback).

Plunge cut—When a router bit plunges, or descends, into a cut. Plunge routers are designed to do this by allowing the router to slide down two tubes until the desired depth of cut is reached.

Pot Life—The length of time two-part glue is usable after it is mixed.

Radius—The distance from the center point to the circumference of a circle.

Set (blade teeth)—The set, or offset, of alternate teeth on a sawblade forms the kerf (the width of the cut), which is slightly wider than the thickness of the blade. This makes cutting circles possible by allowing a band-saw blade some extra room in which to pivot slightly.

Set-up time—The length of time glue takes to bond the surfaces together. This does not mean that the bond has reached its full strength, just that the curing process has started.

Shelf Life—The length of time glue can be stored and still be usable.

Sliding T-Bevel—A very useful tool for transferring angles. It is commonly used to transfer an angle from a drawing to a setting on a saw. It's sometimes called a bevel square.

Stave—An individual piece in a coopered project.

Table of Offsets—A table containing coordinates to locate dots on a grid pattern. By connecting these dots, curved lines can be drawn.

Tangent—A line is tangent to a circle if it intersects the circumference in a single point. In the layout section (Part 1), tangents are used to draw corbel shapes.

Tension—The stretching force on the convex side of a bend. Tension is the common culprit when a bend fails. The tension on the wood pulls the fibers apart.

Trammel points—Trammel points are small clamp devices that fit on a thin board of any length. They're used to draw circles and work like a large compass.

APPENDIX 3:

Sources of Supply

THE AUTHOR'S WEBSITE

Information, updates, videos, workshops, Q&A:

www.woodbender.com

Feel free to e-mail the author:

Lon@woodbender.com

GLUE

Epoxies:

West System Epoxy
Gougeon Brothers, Inc.
PO Box 908
Bay City, MI 48707
517-684-7286
www.westsystem.com

System Three Resins
PO Box 70436
Seattle, WA 98207
800-333-5514
206-782-7979
www.systemthree.com

MAS Epoxies
Phoenix Resins, Inc.
2615 River Rd. #3A
Cinnaminson, NJ 08077
888-637-3769
www.masepoxies.com

Plastic Resin:

Unibond 800
Vacuum Pressing Systems
553 River Rd.
Brunswick, Maine 04011
207-725-0935

Weldwood Plastic Resin Glue
DAP Inc.
Dayton, OH 45401

Bondo:

Bondo All Purpose Household Putty
Dynatron/Bondo Corp.
Atlanta, GA 30331
800-421-2633

VENEER

Austin Hardwoods
2533 Main St.
Santa Ana, CA
714-641-2833

Certainly Wood
1300 Route 78
East Aurora, NY 14052
716-655-0206
www.certainlywood.com

PREMANUFACTURED PLYWOOD SHAPES

Anderson International Trading
(Plywood cylinders and other shapes, bending plywood, baltic birch)
1171 N. Tustin Ave.
Anaheim, CA 92807
800-454-6280
www.ait.wood.com

Keller Products, Inc.
(Plywood cylinders)
PO Box 4105
41 Union St.
Manchester, NH 03108
603-627-7887

RADIUS-CUTTING JIG WITH SUCTION CUP FOR ROUTER

Micro Fence
11100 Cumpston St. #35
North Hollywood, CA 91601
818-766-4367
www.microfence.com

WOODWORKING TOOLS & ACCESSORIES

Fein Power Tools, Inc.
(Sanders, saws, scraping tools, vac systems)
1030 Alcon St.
Pittsburg, PA 15220
800-441-9878

Festo Tools
(Sanders, saws, routers, vac systems)
Toolguide
888-337-8600
www.toolguide.net

The Peck Tool Company
(Radius-cutting hand planes)
PO Box 4744
Boulder, CO 80306
www.PeckTool.com

Lee Valley and Veritas Tools LTD
(Tools, compression straps, steam kettles)
12 E. River St.
Ogdensburg, NY 13669
800-871-8158
www.leevalley.com

Rockler
4365 Willow Dr.
Medina, MN 55340
800-279-4441
www.rockler.com

Woodcraft Supply
PO Box 1686
Parkersburg, WV 26102
800-225-1153
www.woodcraft.com

Index